THERAPEUTIC APPROACHES IN COMMON OBSTETRICS AND GYNECOLOGICAL DISORDERS

VITAMIN D AS PROGESTERONE-LIKE HORMONE

NOVEL APPLICATIONS IN OBSTETRICS AND GYNECOLOGY

Therapeutic Approaches in Common Obstetrics and Gynecological Disorders

Vittorio Unfer - Series Editor

A.G.UN.CO Obstetrics and Gynecology Center, Rome, Italy

Vitamin D as Progesterone-like Hormone Novel Applications in Obstetrics and Gynecology
Vittorio Unfer (Editor)
2020. ISBN: 978-1-53617-516-5 (Hardcover)
2020. ISBN: 978-1-53617-517-2 (e-Book)

Inositols: Guide to Practical Applications in Obstetrics and Gynecology
Vittorio Unfer (Editor)
2019. ISBN: 978-1-53614-912-8 (Hardcover)
2019. ISBN: 978-1-53614-913-5 (e-Book)

Alpha Lipoic Acid: New Perspectives and Clinical Use in Obstetrics and Gynecology
Vittorio Unfer (Editor)
2018. ISBN: 978-1-53613-592-3 (Hardcover)
2018. ISBN: 978-1-53613-593-0 (e-Book)

THERAPEUTIC APPROACHES IN COMMON OBSTETRICS AND GYNECOLOGICAL DISORDERS

VITAMIN D AS PROGESTERONE-LIKE HORMONE

NOVEL APPLICATIONS IN OBSTETRICS AND GYNECOLOGY

VITTORIO UNFER
EDITOR

nova
Medicine & Health
New York

Library of Congress Cataloging-in-Publication Data

ISBN: 978-1-53617-516-5
Library of Congress Control Number: 2020932296

Published by Nova Science Publishers, Inc. † New York

This book summarizes in a transparent manner the most recent knowledge about the role of vitamin D in reproductive medicine. The authors focus, in individual chapters, on the explanation of the role played by vitamin D during the consecutive stages of pregnancy starting from conception. It is an excellent and innovative source of knowledge for anyone assisting patients with reproductive problems.

Artur Wdowiak Associate Professor, MD, Ph.D.
Head of Diagnostic Techniques Unit, Medical University of Lublin, Poland

This is a first review concerning the comparative analysis of the data about progesterone-like activity of Vitamin D in women. The book might help to understand a gap between the relationship of Vitamin D and progesterone physiological functions in human reproduction. Finally, I believe that this book gives an extensive biologic plausibility to the theory that vitamin D levels would be an effective target for fertility improvement and pregnancy outcomes. It appears that the chapter on vitamin D and its progesterone-like action for the general population may have been written based on the most recent trials. However, many questions still remain and will likely fuel ongoing investigation and debate. Are there benefits to vitamin D supplementation regarding fertility outcomes? What is the true benefit on pregnancy outcomes for vitamin D administration in the women population? These questions and others in both fertility and pregnancy will likely keep the book open on vitamin D for the foreseeable future.

Julia Fedotova, Professor, Ph.D.
Department of Chemistry and Molecular Biology, ITMO University, St. Petersburg, Russia.

To date, Vitamin D is fully considered a secosteroid hormone, since it acts on several cellular mechanisms through both genomic and non-genomic pathways. Indeed, accumulating evidence suggests that both calcitriol and, its direct precursor, calcidiol, play a pivotal role to maintain metabolic, hormonal and immune homeostasis in different tissues and microenvironments in humans. In addition, several pieces of evidence support the paramount importance of Vitamin D compounds in human reproduction, both for male and female fertility as well as for several key events such as decidualization, oocyte developmental competence, embryo implantation, placental framework formation, and immune regulation at the maternal-fetal interface. In this scenario, this new book provides a comprehensive summary of the complex mechanisms orchestrated by Vitamin D, from the basic science to its clinical applications. I think that this book could be considered a real milestone for gynecologists, obstetricians, endocrinologists, diabetologists and experts of reproductive medicine, since it can be truly helpful in guiding decision-making processes in clinical practices through an evidence-based approach.

Antonio Simone Laganà, M.D., Ph.D.
Department of Obstetrics and Gynecology, "Filippo Del Ponte" Hospital,
University of Insubria, Varese, Italy.

CONTENTS

FOREWORD

As Endocrinologist, I am deeply involved in hormones and Vitamin D is now considered a full-title hormone, with a wide range of effects on the entire organism. One thing for all: Vitamin D Receptors (VDR) are more diffused than we thought to date; in fact, almost every tissue has VDR.

One of the most intriguing and new fields in which Vitamin D is involved is Obstetrics and Gynecology and this book entitled "Vitamin D as progesterone-like hormone (Novel applications in Obstetrics and Gynecology), is able to fill the gap between the "old" and the "new". In fact, the role of Vitamin D in female reproduction is a recent achievement, that is well described in the first chapter by Unfer and Monastra. The Authors guide the reader in a fascinating journey through the different general aspects of the Vitamin D story, starting from the ancient history to the present days, therefore naturally introducing to the next chapters. Importance is given to VDR that have gained more and more attention in the recent period. In addition, it is well underlined the Progesterone-like activity of Vitamin D and its importance in "providing support for the establishment and the satisfactory development of pregnancy, from the luteal phase of the menstrual cycle onward", (as the Authors stated).

In the second chapter, another hot topic is considered. I am talking about the connection between Vitamin D (and the equally important VDR) and the immune system, with an eye well-opened on the fascinating and very actual field of microbiota. The Author describes different pathologies of the immune system, with a particular accent on Type-1 Diabetes Mellitus prevention and treatment with Vitamin D.

The book proceeds with a chapter on the role of Vitamin D in embryo adhesion and implantation, in which the usefulness of adequate Vitamin D concentrations is a pre-requisite to successfully modulate the mechanism regulating endometrium adequacy and implantation window, in addition to the immunomodulation, that is necessary to tolerate the blastocyst.

Once pregnancy is obtained, it is important to prevent miscarriage and preterm birth. In this view, Vitamin D seems to have a pivotal role in ameliorating expectations. In order to deeply investigate this issue, the Authors of the fourth chapter at the beginning describe the Vitamin D role in pregnancy, underlining that Vitamin D "is one of the principal sites for extra-renal Vitamin D synthesis, with both the maternal and the fetal sides of the placenta cooperating to maintain high localized tissue levels of the hormone." In the second part of their work Menichini and Facchinetti deal with pregnancy loss and preterm birth, concluding that there must be a perfect balance among all the component of the complex mechanism of adaptation between mother and fetus. In this view, Vitamin D plays a role as antibacterial immune response promoter, while suppressing adverse inflammatory adaptive immunity. Vitamin D deficiency affects these processes.

Sometimes, whilst numerous natural attempts, pregnancy does not occur. In this case, Assisted Reproductive Technology (ART) may be of help. The sixth chapter gives a series of information in support of the role of Vitamin D in ART, with a special focus on polycystic ovary syndrome (PCOS) patients in which Vitamin D deficiency seems to be three times less likely to carry on a pregnancy, and that supplementation may significantly revert this outcome. The Authors conclude that there are strong indications in supporting beneficial effects of Vitamin D supplementation, especially in PCOS women in which a special attention should be given when dose finding is concerned.

Another interesting topic addressed in this book is the action of natural molecules, such as vitamin D, magnesium and lipoic acid in controlling the uterine contractile activity. These molecules, when combined, might be useful to treat and/or inhibit uterine contraction, thus preventing preterm birth.

Leaving Obstetrics and jumping to Gynecological problems, the next chapter deals with uterine fibroids and the possible role of Vitamin D, eventually associated with Epigallocatechin Gallate. This is an underestimated pathology that is more frequent than what has been thought, reaching up to 77% of women. Until now, surgery and pharmacological treatments have been the main approaches, however in recent years, non-surgical and non-pharmacological strategies have been developed. Among them, naturally speaking, Vitamin D has demonstrated a role in ameliorating uterine fibroids. It is very interesting to note that, in line with that, Epigallocatechin Gallate has a similar role in the growth of uterine fibroids. Therefore, Porcaro and Orrù, interestingly suggest the combined use of both compounds, as an innovative non-surgical therapy of uterine fibroids.

The group of Modena, in association with Houston, wrote an interesting chapter on the role of Vitamin D, Inositol and Melatonin as support of the luteal phase, which is frequently altered in a bunch of conditions. The Authors extensively reviewed the available literature and concluded that the combination of Vitamin D, Inositol and Melatonin is beneficial to luteal phase deficiency, even though more clinical work is needed to quantify the exact dosage of each component.

Last but not least, the Bizzarri team addressed an issue on the various mechanisms by which vitamin D influences the natural history of cancer. Actions that may influence the induction of apoptosis, stimulation of cell differentiation, anti-inflammatory and antiproliferative effects, as well as invasion and metastasis. They highlight the far-reaching basis for the potential use of Vitamin D in cancer therapy and chemoprevention.

From what above, this book can be easily considered up to date, presenting the last information about Vitamin D in the female universe. Each chapter has been designed to conduct the reader through the topic and, consequently, to the next chapter. Besides that, the contributions from international scientists elevate the global level of the book that is, at one time easy to read, but also contains complex concepts and novel approaches to modern medicine. In this view, thanks to all Authors that made a great effort in order to make complicated issues easy and understandable for the reader that is not an expert of the field.

For all those reasons, I surely suggest this book, as a main reading for scientists as well as general practitioners, who will find innovative and interesting issues to translate in their everyday practice.

Maurizio Nordio, M.D., Ph.D.
Professor of Neuroendocrinology
University “Sapienza”
Rome, December 5th, 2019

PREFACE

This book addresses a very interesting topic and offers a novel insight of the well-known Vitamin D. In the past, this molecule has been wrongly considered from health professionals and researchers a “Vitamin”. The scientific community has now extensively accepted that calcitriol, commonly called “Vitamin” D, is a steroid (more precisely, a secosteroid) hormone. In this book series, the third volume titled “Vitamin D as Progesterone-like Hormone - Novel Applications in Obstetrics and Gynecology”, offers an innovative overview of the critical role played by Vitamin D in the therapeutic approach to common obstetric and gynecological disorders. The nuclear receptors of steroid hormones have common structure/function domains called “nuclear receptors superfamily”. Progesterone and Vitamin D receptors belong to two different classes of this superfamily. Throughout each chapter, we need to keep in mind this relationship as the core of our book.

Vitamin D exerts various and distinct physiological activities during gestation along with progesterone; in some cases, it may act also as its substitute or they may strengthen reciprocally. Like progesterone, Vitamin D seems extremely functional during the luteal phase onwards. A lot of research needs to be done in order to further investigate this novel fascinating role of Vitamin D. We can see the multiple activities also in combination with other molecules, such as alpha-lipoic acid and magnesium for controlling the uterine contractile activity, inositol and melatonin to support the luteal phase or epigallocatechin gallate in the management of uterine myomas. All these attractive insights will be discussed in the present book, analyzing in depth each aspect of this powerful molecule, bringing the instructions from this guide to everyday practice.

ACKNOWLEDGMENT

Writing a book is quite hard, and none of this would have been possible without the great contribution of all Authors. All of them brought their background and experience in this specific field, and showed the diversified roles of Vitamin D, as the main protagonist in this book. A special acknowledge goes to the painter of the image of the book cover. This figure was captured from an original painting "The origin of life" - owned by Prof. Vittorio Unfer - realized in 1981 by Marino Giordanella (one of the 100 painters of Via Margutta). And it is so astonishing how, with a little imagination, it is easy to extrapolate from this image the core of this book and the conceptual expression of Vitamin D. The artist narrates to have no knowledge of physiology and anatomy of the reproductive system and the painting was reproduced during a winter night. The image illustrates the pelvic area at the bottom, the ovaries in the lower left side with eggs (one of which in ovulation); going up to the left, the uterus from which a pregnancy can be glimpsed. And finally, the birth is represented with 2 faces, the mother and the son, looking and confronting each other for the first time.

In: Vitamin D as Progesterone-Like Hormone ISBN: 978-1-53617-516-5
Editor: Vittorio Unfer

Chapter 1

EVOLUTION AND PHYSIOLOGICAL ROLE OF "VITAMIN" D; HORMONAL NATURE AND PROGESTERONE-LIKE ACTIVITY IN PREGNANCY: AN OVERVIEW

Vittorio Unfer*[*]*, MD and Giovanni Monastra
Department of Experimental Medicine, Sapienza University,
Rome, Italy

ABSTRACT

The scientific community agrees that calcitriol ($1,25(OH)_2D$) and calcidiol (25(OH)D) are steroid hormones (more precisely, secosteroids), and the term "vitamin" is outdated when referred to these two molecules. Herein, we adopted the definition "hormone D" either for $1,25(OH)_2D$ and for 25(OH)D, its direct precursor. Both species are biologically active, although $1,25(OH)_2D$ is the most relevant. Conversely, we used the collective definition of "compounds D" (and in a few cases "vitamin" D) for cholecalciferol and ergocalciferol, including the exogenous dietary supplements. Some of these molecules appeared early in the evolution of

* Corresponding Author's Email: vunfer@gmail.com.

life as products of photochemical reactions of ergosterol with ultraviolet B radiation (UVB). They are thought to have protected living organisms against UVB radiations and against microbes. Moreover, they maintained the physiological levels of calcium, necessary for various metabolic functions beside bone formation. Of note, the other two molecules involved in the complex activities carried out by hormone D, namely its receptor and its binding protein, appeared separately during evolution. They played independent roles in the protection and defence of living organisms, and only later became part of the hormone D system. Among the several biological activities of secosteroid hormones, we focused our attention on the effects on menstrual cycle, and on the interplay with progesterone from the early stages of pregnancy until delivery. Many preclinical and clinical studies demonstrated that hormones D and progesterone share several biological activities, through complementary pathways, and collaborate from the luteal phase onward. At the beginning of pregnancy, progesterone and hormones D have similar concentrations, favouring the endometrial receptivity and a successful implantation. The supplementation with hormones D improves fertility and wards off negative outcomes during pregnancy. In conclusion, available data indicate that hormones D have a progesterone-like role in human reproduction and hence physicians should carefully consider the appropriate timing for starting its administration.

Keywords: calcidiol, calcitriol, CYP24A1, CYP27A1, CYP27B1, embryo implantation, inflammation, innate and acquired immunity, in vitro fertilization, pregnancy, progesterone, vitamin D, vitamin D binding protein, vitamin D receptor, xenobiotics

INTRODUCTION: FROM "VITAMIN" D TO HORMONE D

The scientific community largely agrees that calcitriol or 1,25-dihydroxycholecalciferol (1,25(OH)$_2$D), usually called "vitamin" D, is a steroid hormone (more precisely, a secosteroid). In many animals, skin exposure to ultraviolet B radiation (UVB) causes the cleavage of the C9-C10 bond in 7-dehydrocholesterol with the opening of the B-ring, leading to 1,25(OH)$_2$D. It is notewhorty that 7-dehydrocholesterol is also the precursor of cholesterol. This molecule and the related sterols (ergosterol in yeast, phytosterols in plants) are considered a hallmark of eukaryotes as they are deeply involved in the evolution of multicellular organisms (Brown and

Galea 2010). Available data suggest that cholesterol developed also to reduce the damage caused by oxygen, when living organisms reached a subaerial environment. According to these evidences, a protective role for cholesterol emerged in organisms undergoing a critical transition in the course of their biological evolution. A similar function can be assigned to 1,25$(OH)_2$D and calcidiol (known also as calcifediol or 25(OH)D), its direct precursor, and to other molecules with similar structure (Brown and Galea 2010). Chemical and structural similarities between secosteroids and cholesterol justify these conclusions.

1,25$(OH)_2$D was referred to as hormone for the first time towards the end of the XX century (Myrtle JF 1970, Haussler MR 1968). In 1971, two research groups identified its chemical structure (Holick, 1971; Norman, 1971). Norman et al. proposed to reclassify the molecule as a steroid hormone, based on the following features: it is secreted by the kidneys; it accumulates in the target intestinal mucosa, exerting its physiological activity on calcium metabolism (Norman 1971). Anthony W. Norman (Norman 2012) reports that the term "vitamin" D originated because of historical reasons: its role in rickets prevention and/or treatment. In fact, a vitamin is normally defined as a molecule essential for animal life, provided from the regular diet in very low amounts and that animals cannot synthesize. Accordingly, animals need a constant external supply of vitamins. As our skin produces "vitamin" D ("vitamin" D3, specifically) by converting cutaneous 7-dehydrocholesterol when exposed to UVB, the term "vitamin" should be more appropriately used only for the exogenous supplementation in case of deficiency, either because of limited exposure to sunlight or UVB sources, or because of specific genetic "defects." Under physiological conditions and normal exposure to sunlight, all mammals produce enough vitamin D3 to meet their needs, without the requirement of a nutritional uptake. This situation perfectly fits with its hormonal nature. Later in this chapter we will discuss other important characteristics that strengthen such classification.

After much debate, only in 1997, the "Institute of Medicine (US) Standing Committee on the Scientific Evaluation of Dietary Reference Intakes" defined 1,25$(OH)_2$D as a "biologically active hormone" (Intakes

1997). Nonetheless, a few scholars still disagree with this definition (Vieth 2004), despite the enormous amount of evidence gathered over the past decades.

In this chapter, we adopted the term "hormone D" to identify $1,25(OH)_2D$ and 25(OH)D, both biologically active, whereas we used the collective definition "compounds D" for cholecalciferol and ergocalciferol, including the exogenous supplements. Where appropriate we used also the old definition "vitamin" (in brackets). All these compounds act within a complex system (the hormone D system) where vitamin D binding protein (DBP) and vitamin D receptor (VDR) are other essential components. The presence of DBP and VDR further supports the use of the term "hormone," instead of "vitamin" D.

When Did These Secosteroids Appear in Evolution?

The first evidence of the appearance of one form of compound D in the course of the biological evolution dates back to more than 500 million years ago (Wacker and Holick 2013, Goring and Koshuchowa 2015, Holick 2011). Although the dating may be questionable, compound D seems to be one of the first, or even the very first hormone that appeared on Earth.

Such earliest form of compound D was probably synthesized by phytoplankton, specifically the coccolithophorid unicellular alga *Emiliania huxleyi*, which features a calcium carbonate exoskeleton. According to Holick, this alga have remained unchanged in the Sargasso Sea (Atlantic Ocean) for more than 500 million years (Wacker and Holick 2013) and produces large quantities of ergosterol, up to 0.1% of its dry matter. When exposed to sunlight, ergosterol absorbs UVB radiation and undergoes a photolytic reaction to generate the so-called previtamin D2, an unstable molecule that quickly evolves into ergocalciferol. The biological function of this energy-consuming system is still unclear. Three hypotheses have been proposed, all related to a "protection" or "warning" role played by secosteroids. The first is based on the efficient absorption of UVB radiation; secosteroids would have acted as a natural sunscreen to protect UVB sensitive macromolecules such as proteins, RNA and DNA. In fact, ergosterol and its photoproducts absorb solar radiation between 280 and 320 nm, the range of wavelengths that typically causes damages to the macromolecules within living cells (Holick 2011). It was speculated that, early in evolution, the first living organisms started to utilize solar energy for photosynthesis, but, at the same time, they needed a protection against sunlight to strongly reduce the damage of sensitive molecules to UVB. Acting together, the different photoproducts were likely very useful to preserve the biological structures (Holick 2003).

The second hypothesis argues that ergosterol and its photoproducts served as sensors for the intensity of incident solar radiation, signalling the excessive radiation and the need for the organism to go deeper in the water and avoid further exposure.

The third hypotesis considers that ergosterol was mainly present in the plasma membrane and included within the lipid bilayer. Ergosterol has a rigid planar structure, which is not particularly suitable for the biological membranes; its transformation into ergocalciferol, a more "elastic" molecule, increased membrane flexibility and selective permeability, intrinsic properties of biological structures. Some authors speculated that this change allowed a pore opening process and the resulting bidirectional passage of ions (Ca^{++}, for example). Allegedly, this represents the first relationship between sun exposure and the homeostasis of Ca^{++} in living organisms (Holick, Tian, and Allen 1995, Holick 2003, Holick et al. 2007).

Moreover, phytoplankton, such as *Emiliania huxleyi*, and zooplankton may have an exoskeleton and need to regulate calcium availability during its formation and resorption. Compounds D are supposed to play a role also in such processes.

New Functions for These Molecules

The connection between 1,25$(OH)_2$D (or other related compounds) and calcium went far beyond the formation and maintenance of these types of primordial defence from the environmental challenges. Natural selection played its role in this evolutionary context. Bikle hypothesized that their function developed along with the evolution of eukaryotic organs to help maintain physiologically different calcium levels outside and inside the cell (for instance, compartmented in mitochondria). This role probably supported essential biological functions since calcium ions are necessary as second messengers for a number of signal transduction pathways (Bikle 2011). The existence of a link between innate immune response and hormone D/compounds D was recently proved. Mainly 1,25$(OH)_2$D exerts antibiotic-like actions by inducing the synthesis of peptides with antimicrobial activity and stimulating the autophagy (Shuler et al. 2013, Vanherwegen, Gysemans, and Mathieu 2017). Hormones/compounds D likely played this role, before the evolutionary novelties of calcified eggs and skeleton appeared (Hernigou, Auregan, and Dubory 2018). Overall, it seems that, due to their chemical structure, 1,25$(OH)_2$D and the related secosteroids protected living beings against external threats by means of different strategies.

Another evolutionary role of compounds D involves their association with Toll-like receptors (TLR) in modulating the innate immune response. TLRs are proteins expressed on sentinel cells (monocytes, macrophages, dendritic cells, mast cells, neutrophils, B and T cells, fibroblasts, endothelial and epithelial cells) and recognize pathogen-associated molecular patterns as well as damage-associated molecular patterns. Of note, TLRs up-regulate the expression of CYP27B1, which converts 25(OH)D into the more active 1,25$(OH)_2$D (Takeyama et al. 1997). 1,25$(OH)_2$D, in turn, stimulates the production of the Cathelicidin Anti-Microbial Peptide (CAMP) (Shuler et al. 2013) and the release of human defensins, small cysteine-rich proteins. CAMP and human defensins act against pathogens like mycobacterium tuberculosis by disrupting the bacterial biofilm and/or stimulating autophagy. Such antimicrobial activity is exerted in the epithelium, in case of injuries, and in the placenta, preventing fetal infections.

During evolution, compounds D progressively became part of several other biological processes. One of these is the regulation of calcium and phosphate absorption to build the skeleton of the organisms that faced the critical transition from marine to terrestrial life, when the environmental conditions drastically changed. Thus, hormone D system became essential in terrestrial vertebrates coping with the challenge of higher gravity and the lack of calcium (Bouillon and Suda 2014). Calcium concentration varies between cellular and extracellular milieu, depending on the site (compartmentalization): close to 100 nM in the overall cell, up to 1 mM in subcellular organelles (i.e., endoplasmic reticulum and mitochondria) and between 1 and 2 mM in extracellular fluids. Therefore, calcium concentration is significantly lower in the intracellular than in the extracellular fluid. In fact, cells cannot be metabolically active and survive out of this gradient. Calcium concentration in the ocean is about 10 mM. Thus, saltwater organisms must face the problem to limit the influx of this ion and/or get rid of its excess. Conversely, organisms living in the subaerial environment need to obtain calcium through their roots or digestive system. In the presence of an exo- or endocalcified skeleton, they have greater demands for calcium over and above the requirements to maintain cellular function (Bikle 2011, Bouillon and Suda 2014, Hernigou, Auregan, and Dubory 2018). On the other hand, phosphate concentration is much lower in seawater than in body fluids. Like calcium, phosphate is vital for a large number of cellular functions such as handling and storing energy, second signaling (phosphorylation of proteins by kinases), and building important structures (DNA/RNA, membranes etc). During the evolution of life in seawater, calcium was plenty available (or in excess) whereas phosphate supply was limited. When "conquering" the earth, amphibians and tetrapods faced an environment characterized by low amounts of calcium and relatively rich in phosphate. In addition, terrestrial animals have to tolerate a six-fold higher gravity in comparison with marine animals. This challenge was overcome by the evolution of a solid bone structure, with maximal strength and minimal weight, to support muscles and total body mobility.

This evolutionary step required a highly efficient system of intestinal calcium absorption where the activity of hormone D was indispensable (Hernigou, Auregan, and Dubory 2018, Bouillon and Suda 2014).

VDR AND DBP IN EVOLUTION

As mentioned before, the fully evolved hormone D system has also other components: the vitamin D receptor (VDR) and the vitamin D binding protein (DBP), an α2-globulin produced in the liver. Before being coupled with hormone D, VDR played an autonomous defensive role for living forms in the early steps of evolution. Interestingly, compounds D and VDR evolved separately and eventually converged in the evolution timeline.

Ancient VDR was a xenobiotic receptor, able to neutralize and degrade biotoxins with sterol ring scaffolds present in aquatic milieu (Bouillon and Suda 2014). Evolutionary change in the vertebrate VDR resulted in the acquisition of new immunological functions, exploiting its previous capability to interact with sterol ring scaffolds, i.e., with molecules similar to $1,25(OH)_2D$. The co-adaptation of this receptor with the acquisition of sterol biosynthesis in vertebrate animals set a stage for repurposing and linking a pre-existing host-protection mechanism from harmful xenobiotics in a new context, becoming an important regulator in immunity. However, VDR still retains some of its ancestral ability in humans, where it is important in eliminating lithocholic acid (LCA), a toxic secondary bile acid present in the colon, by activating the CYP3A4 P450 enzyme (Newmark, Dantoft, and Ghazal 2017).

Finally, in order to provide a better understanding of the evolutionary origin of DBP, it is worth mentioning the functions indirectly related to cholecalciferol, 25(OH)D and $1,25(OH)_2D$. DBP is a serum protein that currently plays immunomodulatory functions in the lungs, predominantly associated to macrophage activation and neutrophil chemotaxis. The evidence indicating this association is related to chronic obstructive pulmonary disease (COPD). Instead, fewer data are available regarding other airways diseases, such as asthma and bronchiectasis. DBP also acts as a scavenger protein to clear extracellular G-actin released from necrotic cells, which may be of relevance in severe lung infections and acute lung injury (Chishimba et al. 2010). Like VDR, DBP probably appeared and evolved as a molecule playing an autonomous defensive role for the organisms; in a certain phase of biological evolution, the three molecules were incorporated into the hormone D system.

Rickets and "Vitamin" D in the History of Medicine

Early reports of bony deformities in infants, likely ascribable to rickets, date back to the second century AD (Rajakumar 2003). The Greek physicians Galen (129 – circa 200/216 AD) and Soranus of Ephesus (1st/2nd century AD) described this condition in children living in the poor and dark neighbourhoods of the imperial Rome. But rickets is truly a modern age pathology (Wharton and Bishop 2003), mainly connected to pollution and novel life habits, which drastically reduced people's exposure to sunlight. The disease became indeed endemic in the South-Western England at the beginning of the industrial revolution (Dunn 1998). In the mid-17th century Whistler and Glisson independently described an unknown ailment, identifying a relationship between age and onset of symptoms. In 1634 the term rickets appeared for the first time as death cause in the London Bills of Mortality (O'Riordan and Bijvoet 2014). Glisson recognized the disease as non-congenital, nor inherited or contagious, and the first investigations on its aetiology suggested a connection between incidence and dietary habits (Dunn 1998).

Remedies to rickets were found two centuries later, when the disease became rampant in polluted cities in Europe and Northern United States, and consisted of sunbaths and dietary animal fat. Analysis of geographical incidence, and empirical observations of anti-rachitic properties of cod-liver oil were crucial factors in curing rickets (Guy 1923). Cod-liver oil is an old remedy of popular medicine, used for a long time to cure several illnesses. Of note, many of them are now treated also with "vitamin" D (Guy 1923).

McCollum later figured out that a fat-soluble component of animal food, different from vitamin A, restores physiological calcium metabolism and bone growth. Three vitamins had already been discovered back then and named after the first letters of the alphabet. With the same logic, McCollum called such new component "vitamin D" (McCollum 1922).

At the beginning of 1900s, experiments on rats demonstrated the influence of irradiated food intake on bone growth (Steenbock 1924), suggesting a connection between light exposure and calcium metabolism (Carpenter and Zhao 1999). But the rationale behind the benefits of sunbaths in treating rickets was elucidated much later (Webb and Holick 1988), when it was discovered that humans endogenously synthesize "vitamin" D from 7-dehydrocholesterol (Wolf 2004).

Deficiency of $1,25(OH)_2D$ and related secosteroids proved to adversely affect calcium metabolism, osteoblastic activity, matrix ossification, bone remodeling and bone density, causing rickets in the developing skeleton as well as osteomalacia and (without a general agreement) osteopenia/osteoporosis in adults, in addition to other similar diseases (Christodoulou et al. 2013).

This bone-focused perspective, persisting until recently, was obviously determined by the circumstances linked to its discovery, which finally allowed to solve the dramatic problem of rickets (Norman 2012, Rajakumar 2003).

Metabolic Pathways of Compounds and Hormones D: Similarities with Steroid Hormones

The general name "vitamin" D included a family of molecules of animal and vegetable origin characterized by similar structures, differing in the aliphatic side chain, with cholecalciferol (D3) and ergocalciferol (D2) being the most biologically relevant analogues (DeLuca 2004). In humans, under normal conditions, the skin produces at least 80% (or more) of the daily needs of cholecalciferol from 7-dehydrocholesterol, in a two-step process that requires an initial photochemically driven ring-opening reaction and a subsequent a thermal isomerization. Human skin can thus be considered an endocrine gland that secretes a pro-hormone, which undergoes two enzymatic hydroxylation reactions, initially yielding 25(OH)D, and eventually $1,25(OH)_2D$ (Figure 1). The former is the most long-lived, abundant form of hormone D in the bloodstream, whereas the latter is the short-lived, most active form.

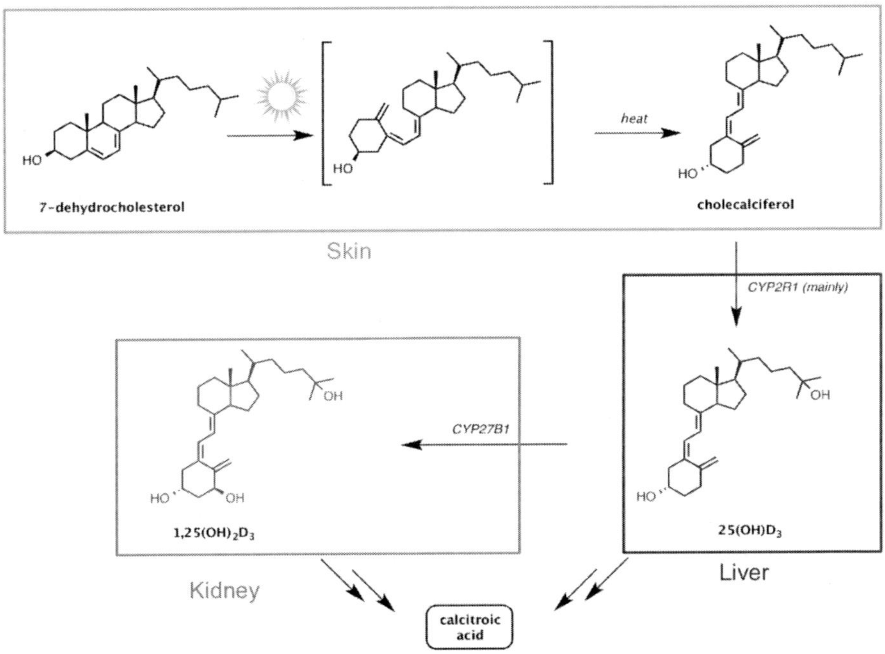

Figure 1. Main metabolic pathway of compounds D3. Sunlight promotes the synthesis of cholecalciferol from 7-dehydrocholesterol in the skin. Cholecalciferol is then converted into $25(OH)D_3$ by CYP2R1 enzyme in the liver, and eventually into $1,25(OH)_2D_3$ by CYP27B1 enzyme in the kidneys. Both $25(OH)D_3$ and $1,25(OH)_2D_3$ can be further oxidized by CYP24A1 enzyme to yield different catabolites, whose major component is calcitroic acid.

Hormone D circulates in the blood stream bound to a protein carrier, a feature typical of the steroid hormones: about 85% is bound to vitamin D binding protein (DBP), which is – as mentioned before – a serum α2-globulin (MW: 52-59 kDa) synthesized in the liver and discovered in 1975 (Speeckaert et al. 2014, Delanghe, Speeckaert, and Speeckaert 2015, Cooke and Haddad 1989); the remainder 15% binds to albumin. Unlike pregnant women, normal subjects have also very small amounts, approximately 0.4% of total 1,25$(OH)_2$D and 0.03% of total 25(OH)D, as unbound, free form (Bikle and Schwartz 2019). Instead, in pregnancy there are time dependent changes in DBP, with almost two-fold increases between the second and the third trimester. Despite these marked DBP changes, mean free 25(OH)D may be the same as, or only slightly lower than non-pregnant women but with less variability (Bikle and Schwartz 2019).

Ergocalciferol, which is of plant origin, presents a slightly lower affinity for DBP than cholecalciferol, its counterpart of animal origin (Tsuprykov et al. 2018).

Hormones are chemical messengers that bind to specific receptors (Norman, Mizwicki, and Norman 2004, Norman 2006, Levin 2015), triggering a signal transmission that usually occurs thanks to the activation of a second messenger pathway. The second messenger, in turn, induces rapid cellular responses that range from seconds (e.g., opening of ion channels) to 10-60 minutes (e.g., activation of phosphatidylinositol-3'-kinase or endothelial nitric oxide synthase). Unlike growth factors and peptide hormones, steroid hormones can cross the cell membrane and interact with nuclear receptors that bind to DNA, activating or repressing gene transcription (O'Malley 1971). This genomic response commonly takes hours to days and may be blocked by transcription or translation inhibitors.

Accordingly, all known actions of hormone D in cells are mediated by VDR. Human VDR is a DNA-binding protein (MW = 50÷60 kDa) with biochemical properties resembling those of receptors for steroid and thyroid hormones (Haussler 1986). VDR is ubiquitously expressed in human tissues (see Table 3) in multiple isoforms (Poon et al. 2012, Wang, Zhu, and DeLuca 2012), likely to tailor the response to compounds D. This complex scenario originally suggested an involvement of hormone D-mediated effects in a spectrum of different contexts besides the physiologic development and maintenance of the skeleton. Like all steroid hormones, hormone D interacts with its receptor located both on the cell membrane (associated with the caveolae) and in the nucleus. As for other hormones, the receptor-hormone D complex can bind to DNA and activate or repress gene transcription.

As repeatedly mentioned in this Chapter, the label "vitamin" for compounds D is now considered simplistic and misleading. All the characteristics listed in Table 1 prompted the majority of researchers to define them as fully-fledged hormones, and the nomenclature should be revised in this perspective.

Table 1. Summary of the features in common between steroid hormones and "vitamin" D

Steroid Hormones	"Vitamin" D
Endogenously produced by glands.	≥ 80% produced by the skin upon exposure to sunlight.
Mostly bound to plasma transport proteins.	Bound to Vitamin D binding protein (DBP), produced in the liver.
Receptors typically located in the cytoplasm and in nucleus of target cells.	Vitamin D receptor (VDR) localized within: caveolae of the plasma membrane; ii) nucleus.
Receptor-hormone complex binds to DNA, activating or repressing gene transcription.	VDR-Vitamin D complex binds to DNA, modifying gene expression.

Although systemic levels of hormones D are maintained by enzymes in the liver and kidneys (and testes in males) (Foresta et al. 2010, Choudhary et al. 2005), all target organs express the enzymes to locally synthesize and degrade both 25(OH)D and 1,25(OH)2D. Local production of hormones D and the discovery of several metabolites having affinity for VDR, greatly complicates outlining the overall regulation of compounds D metabolism. It comprises an elaborate network of feedback systems that tune specific and diverse activities in different tissues, under various conditions. Specific cytochrome P450 (CYP) enzymes constitute the machinery that controls the biosynthesis and the activity of hormones D (Prosser and Jones 2004). Microsomal CYP2R1 is the enzyme that produces most of the circulating 25(OH)D (Zhu et al. 2013). Mitochondrial CYP27A1, originally thought to be the primary 25OHase in this process, later proved to be one of several enzymes that provide a backup system to produce 25(OH)D in case of impaired CYP2R1 pathway (Thacher and Levine 2017). CYP27B1 is the only known human 1-alpha-OHase, able to convert 25(OH)D into the more active 1,25$(OH)_2$D. Although 25(OH)D is less active than 1,25$(OH)_2$D, it still binds to VDR and exhibits a biological activity. (Takeyama et al. 1997, Tuohimaa et al. 2013, Fawaz et al. 2016, Wang, Song, and Ma 2018).

A single hydroxylase, CYP24A1 (Jones, Prosser, and Kaufmann 2012), catalyzes all the reactions to degrade both forms of hormones D to a major catabolite, calcitroic acid – a hydrophilic compound, having a low affinity for VDR, which is eventually excreted as bile into the intestine (Yu and Arnold 2016).

The issue of compounds D metabolism is still a debated topic, subjected to continuous "rethinking," as demonstrated by the studies of Tuckey et al. (Tuckey, Cheng, and Slominski 2019) that reported about 20S-hydroxyvitamin D3 (20S(OH)D3), which is biologically active and has several analogies with 1,25$(OH)_2$D, but devoid of any calcemic effect. This field of research is expected to provide many fascinating discoveries in the next future.

In an oversimplified view – functional to the purpose of this chapter – homeostasis of hormones D, especially the concentration of 1,25(OH)2D, is tightly regulated through the control of the expression and the activity of CYP27B1 and 24A1. In response to low plasma calcium concentration, parathormone (PTH) induces the expression of CYP27B1 and, hence, the biosynthesis of 1,25(OH)2D. 1,25(OH)2D triggers a negative feedback that inhibits the activity of CYP27B1, halting its own synthesis. At the same time, increased concentrations of 1,25$(OH)_2$D upregulates the expression of CYP24A1, promoting its catabolism and the degradation of 25(OH)D. 1,25$(OH)_2$D also induces the expression of the Fibroblast Growth Factor (FGF)-23, which contributes to control the concentration of 1,25$(OH)_2$D by down-regulating PTH and CYP27B1, and by up-regulating CYP24A1 (Petkovich M 2011).

COMPOUNDS D: INTESTINAL ABSORPTION AND STORAGE

In healthy subjects, compounds D are absorbed at the intestinal level, with an average efficiency of 78% (ranging between 55% and 99%). The main absorption site in human gut is unknown, but uptake in rats occurs both in the jejunum and in the ileum. Low-hydroxylated forms (ergocalciferol and cholecalciferol) cross the intestinal barrier by passive diffusion and/or by active transport, mediated by specific proteins: SR-B1 (Scavenger Receptor class B type 1), CD36 (Cluster Determinant 36) and NPC1L1 (Niemann-Pick C1-Like 1). Such proteins are also involved in the uptake of cholesterol, vitamin E (Reboul et al. 2006, Narushima et al. 2008), and carotenoids (Reboul et al. 2005, van Bennekum et al. 2005, Moussa et al. 2008). "Vitamin" D uptake is likely protein-mediated at low concentrations (e.g., dietary intake), whereas passive absorption becomes prevalent at high concentrations (e.g., when administered as supplement or drug – Table 2).

Table 2. Vitamin D - Maximum intake for supplementation (1 μg = 40 IU)

Geographic Areas	Dose per day	Dose per month
UE (Italy)	50 μg/day	1500 μg/month
USA	100 μg/day	3000 μg/month
South East Asian Nations	25 μg/day	750 μg/month

Once inside enterocytes, "vitamin" D is transferred to the endoplasmic reticulum, then to chylomicrons and finally reaches the lymphatic pathway. Conversely, high-hydroxylated forms (25(OH)D and 1,25(OH)$_2$D) are absorbed three times more efficiently and reach the portal vein pathway (Borel, Caillaud, and Cano 2015).

Due to their lipophilic nature, compounds D are mainly stored in muscle, liver and adipose tissues (Mawer et al. 1972). This characteristic is also responsible for their slow turnover in the body (half-life ~2 months). The physiologic concentrations of 25(OH)D in the blood serum are between 75 and 100 nmol/L (30-40 ng/ml), whereas levels between 50-75 nmol/L (20-30 ng/ml) insufficiency and levels below 50 nmol/L (20 ng/ml) indicate deficiency. The half-life of 25(OH)D is ~15 d, or even more, whereas cholecalciferol and 1,25$(OH)_2$D have a much shorter half-life (Table 3) (Jones 2008). Due to its stability and high concentration, 25(OH)D is the most reliable marker for evaluating the levels of hormones D in the body.

Table 3. Half-life of compounds D (Jones 2008, Haddad and Rojanasathit 1976, Levine et al. 1985)

Compound D	Half-life
cholecalciferol	24 hours
25(OH)D	15-20 days
1,25$(OH)_2$D	4-15 hours

Studies on "vitamin" D levels in plasma highlighted that serum peak of cholecalciferol appears the day after administration; on the other hand, 25(OH)D peaks only 7 days after administration. Such delay is likely ascribable to the metabolic conversion in the liver (Heaney et al. 2008).

Factors That May Influence "Vitamin" D/Hormone D Absorption

Food

Food matrices have apparently little to negligible effect, but the area has scarcely been investigated and further in-depth studies are desirable (Borel, Caillaud, and Cano 2015).

Drugs

Many common drugs may interfere with the absorption of these secosteroids and/or subsequent metabolism, reducing the circulating levels of hormones D. They include: bile acid sequestrants (colestipol and cholestyramine), lipase inhibitors (orlistat), statins, antimicrobials (rifampin, isoniazid, hydroxychloroquine), antiepileptic drugs, corticosteroids, immunosuppressive agents (cyclosporine, tacrolimus), chemotherapeutic agents (etoposide, epipodophyllotoxin, cyclosphosphamide, ifosfamide, vincristine, vinblastine, paclitaxel, docetaxel, irinotecan, tamoxifen and imatinib), highly active antiretroviral agents, histamine H2-receptor antagonists (Robien et al. 2013).

Smoking

Smoking may lead to depleted levels of circulating hormones D. Decreased circulating 25(OH)D, up to 10%, was reported for perimenopausal smokers (Brot, Jensen, and Sorensen 1997), with a positive correlation with smoking pack-years (Piazzolla et al. 2017). Analogous effect (10-15% decrease) was observed in pregnant smokers. Unlike 1,25$(OH)_2$D, 25(OH)D can cross the placenta (Larque et al. 2018) and depleted maternal 25(OH)D levels may result in a low fetal hormone D status.

From Skeleton Diseases to the New Therapeutic Sectors

Besides its involvement in the normal growth of the skeleton and in the related diseases, 1,25$(OH)_2$D is an essential molecule for the physiological activity of many tissues and organs. A huge amount of preclinical researches and clinical studies demonstrated that 1,25$(OH)_2$D and other hormones D play several physiological key roles in different areas and, consequently, their deficiency causes detrimental effects on our health (Wimalawansa 2019, Wang et al. 2017, Monastra et al. 2018, Berridge 2017, Di Somma et al. 2017, Kulling et al. 2017, Holick 2017, Marino and Misra 2019, Lips 2006, Shuler et al. 2013, Vanherwegen, Gysemans, and Mathieu 2017). Moreover, thanks to its anti-inflammatory, antiproliferative, anti-angiogenic and antibacterial activity,"vitamin" D administration proved effective in treating several systemic illnesses in the fields of endocrinology, immunology, oncology, rheumatology, hematology, neurology, dermatology, and obstetrics/reproductive medicine (Wang et al. 2017,

Monastra et al. 2018, Harrison et al. 2019, Kulling et al. 2017, Holick 2017, Marino and Misra 2019, Lips 2006, Shuler et al. 2013, Vanherwegen, Gysemans, and Mathieu 2017, Sassi, Tamone, and D'Amelio 2018, Liu et al. 2018, Krishnan and Feldman 2011, Feldman et al. 2014). The latter, in particular, is the therapeutic area more in-depth discussed in this Chapter.

Secosteroids in Reproduction

In the mid-1930s a group of researchers reported the hormonal properties of a group of molecules with cholesterol-like structure, a seventeen-carbon cyclopentanoperhydrophenanthrene scaffold. While carrying out experimental studies with animals in the field of reproduction, they observed that injection of each of these compounds, dissolved in sesame oil, induced oestrus in ovariectomized rats and that the classic steroid structure was responsible for such effect. In particular, ergosterol and its irradiation product (calciferol, crystalline vitamin D) exhibited marked activity, and neo-ergosterol had even greater activity (Cook J. W. 1934).

Further investigations highlighted that ergosterol and 7-dehydrocholesterol act synergistically with progesterone in sustaining pregnancy (Feyel-Cabanes 1949) or in obtaining implantation in ovariectomized rabbits (Chambon 1951), when administered at non-estrogenic, nor progestogenic doses.

For quite some time these observations were disregarded, and reconsidered only between the end of 1960s and the beginning of 1970s, when new studies shed light on activities of $1,25(OH)_2D$ that go beyond bone growth and rickets treatment. Around the same time, the structure of $1,25(OH)_2D$ was elucidated and its nuclear receptor (VDR) identified (Haussler MR 1969, Tsai HC 1973), leading to define “vitamin” D for the first time as a steroid hormone (Haussler MR 1968, Myrtle JF 1970) – precisely a secosteroid.

COMPOUNDS D, DBP AND VDR

Sharing a common ancestor, the structures of compounds D are closely related to corticoids and sex hormones like testosterone, progesterone and estrogens. In sunny areas, endogenous production of cholecalciferol meets the daily requirements in healthy young subjects, but dark complexion decreases sunlight absorption. Blacks and Hispanics in the US, for example, generally lack adequate levels of endogenous cholecalciferol, outlining an ethnic predisposition for low "vitamin" D status. Other physical conditions represent significant and independent risk factors for the reduced concentration of circulating hormones D. These include obesity, hypertension and low levels of high-density lipoprotein (Forrest and Stuhldreher 2011).

As mentioned, steroid hormones bind to transport proteins in plasma and, according to the free hormone hypothesis (Siiteri et al. 1982), only the small unbound fraction reaches its receptors in target cells. Likewise, hormones D largely bind to DBP and to albumin. Some authors suggest that the free hormone hypothesis applies also in this case, but the question is still debated. Compounds D bound to DBP, in fact, can cross the cell membranes of specific organs and tissues (namely placenta, parathyroid and renal proximal convoluted tubules) through endocytosis by the cubilin/megalin complex (Bikle, Malmstroem, and Schwartz 2017), locally increasing the concentration of available hormone.

Hormones D, as steroid hormones (see above), exert either genomic and nongenomic actions (Norman 2006): interaction with VDR on cell membrane induces the formation of second messengers (cAMP, diacylglycerol, inositol triphosphate arachidonic acid) or protein phosphorylation; interaction with nuclear VDR may initiate a genetic response (Haussler et al. 2011). In the nucleus, bound VDR forms a heterodimer with the Retinoid-X-Receptor (RXR). VDR/RXR subsequently binds to DNA regions known as Vitamin D Responsive Elements (VDREs) and modifies the expression of several genes by recruiting numerous coactivators or corepressors (Nezbedova and Brtko 2004, Darwish and DeLuca 1993, Carlberg, Seuter, and Heikkinen 2012).

The characteristics of the receptor's activity, along with the discovery of many enzymes responsible for the metabolism of this group of molecules, strongly suggest implications of hormones D in several metabolic and regulation processes (Mizwicki et al. 2010). In addition, VDR is almost ubiquitously distributed in human tissues and organs, including ovarian, myometrial and endometrial cells, and the placenta (Table 4) (Valdivielso 2009, Christakos et al. 2013, Acker et al. 1982, Delvin et al. 1985, Kachkache et al. 1993).

Such considerable expression of VDR in the female reproductive system, along with the timing in $1,25(OH)_2D$ regulation, highlights the involvement of low hormone D status in fertility and in the onset of pregnancy complications.

Table 4. Human tissues expressing VDR

Adipose	**Kidney**	**Pituitary**
Adrenal	Liver (fetal)	Placenta
Bone	Lung	Prostate
Bone marrow	Lymphocytes (B & T)	Retina
Brain	Muscle, embryonic	Skin
Breast	Muscle, smooth	Stomach
Cartilage	Osteoblast	Testis (and spermatozoa)
Colon	Ovary	Thymus
Epididymis	Pancreas β cell	Thyroid
Hair follicle	Parathyroid	Uterus (myometrium and endometrium)
Intestine	Parotid	Muscle, cardiac

Modified from: Norman AW. From vitamin D to hormone D: fundamentals of the vitamin D endocrine system essential for good health, Am J Clin Nutr. 2008 Aug; 88(2):491S-499S.

PROGESTERONE IN THE LUTEAL PHASE

The regulation of the female reproductive system involves the interplay of many actors, whose concentration must reach appropriate levels in specific timeframes. In particular, progesterone and $1,25(OH)_2D$ are essential for a successful gestation, and their metabolism is tightly controlled throughout the menstrual cycle.

Menstrual cycle alternates two main stages: 1) follicular or proliferative phase (1st to 14th day of the menstrual cycle); 2) luteal or secretive phase (15th to 28th day of the menstrual cycle, normally). In the first, the follicles inside the ovary develop to prepare for ovulation. In the second, the corpus luteum (CL), a transient endocrine gland that forms at this stage, secretes progesterone (hence secretory phase). CL undergoes rapid neovascularization under the influence of the vascular endothelial growth factor (VEGF) and the fibroblast growth factor (FGF), both upregulated in the luteinized granulosa cells and critical for embryo implantation (Chakraborty, Das, and Dey 1995, Sak et al. 2013). In response to low stimulation of the luteinizing hormone (LH) and appropriate levels of cholesterol, luteal cells synthesize progesterone in a pulsatile way, resulting in a 7-fold concentration change within a few hours. In the luteal phase, progesterone inhibits proliferation of endometrium and induces a receptive state for the blastocyst implantation, maintaining early pregnancy (Young 2013).

In this scenario, hormones D modulate the innate and adaptive immune system, control cell proliferation and regulate metabolic processes, affecting embryo implantation, cytokine production and immune response to infections (Niswender et al. 2000, Mesen and Young 2015, Herr, Bekes, and Wulff 2013, Davis and Rueda 2002).

LEVELS AND MUTUAL ACTIVITY OF HORMONES D AND PROGESTERONE IN EARLY PREGNANCY AND SUBSEQUENT STAGES

Hormones D

1,25$(OH)_2$D metabolism heavily changes during menstrual cycle and pregnancy, with pregnant women exhibiting higher values of circulating hormones D, independently of calcium homeostasis (Kumar 1980, Hartwell, Riis, and Christiansen 1990, Salle et al. 2000). As mentioned, the receptor

VDR is present in the ovaries, uterus and placenta; it is expressed in endometrial stromal cells throughout the entire menstrual cycle and in early pregnant decidual cells (Vigano et al. 2006), and its number constantly increases in the first and the second trimester of pregnancy. Correspondingly, the levels of CYP27B1 mRNA remain constant in endometrial stromal cells throughout the menstrual cycle, but they significantly increase in decidual cells (Vigano et al. 2006). Remarkably, hormones D can be detected also in the seminal fluid (Blomberg Jensen 2014) and support a successful onset of pregnancy (Boisen et al. 2017, Lorenzen et al. 2017). This widespread co-expression of VDR and CYP27B1 (Hewison et al. 2000) provides the basis for an autocrine/paracrine action of hormones D, modulating both acquired and innate immunity at the fetal-maternal interface – specifically the inflammatory response that occurs in the initial stage of pregnancy (Evans et al. 2006). In the implantation process, blastocysts produce pro-inflammatory cytokine IL-1β, which stimulates endometrial dendritic cells and macrophages to express CYP27B1, and consequently increases $1,25(OH)_2D$ concentration (Vigano et al. 2006). Tumor necrosis factor (TNF-α), the other main pro-inflammatory cytokine, is much less effective in this process. In vitro experiments proved that 1,25(OH)2D inhibits the proliferation of T helper 1 (Th1) cells, decreasing the production of the associated cytokines (IFN-γ, IL-2, TNF-α); at the same time, it stimulates the secretion of T helper 2 (Th2) cytokines, including IL-10 (Figure 2) (Ota et al. 2015). Furthermore, 1,25(OH)2D significantly decreases the cytotoxicity of natural killer (NK) cells in a dose-dependent manner, suppresses the secretion of pro-inflammatory cytokines (IFN-γ and TNF-α) and increases the release of IL-10, CSF2, IL-1 and IL-6 (Ota et al. 2015), modulating the inflammatory response that physiologically develops at the initial stage of pregnancy (Evans et al. 2006). In patients with recurrent miscarriages, supplementation with vitamin D significantly lowers the frequency of T helper 17 (Th17) cells, decreasing the Th17/Treg ratio in peripheral blood and reducing the inflammatory state (Rafiee et al. 2015). Analogous effect was observed in the induction and upregulation of the suppressive activity of T regulatory (Treg) cells on dermal inflammation in patients with skin-draining lymph nodes (Gorman et al. 2016).

On the endometrium, binding of hormones D to VDR upregulates the expression of Calbindin and Osteopontin (Vigano et al. 2006), essential proteins for embryo implantation, placenta development and ultimately for successful pregnancy (Vigano et al. 2006, Choi, Leung, and Jeung 2005, Emam, Abouelroos, and Gad 2016, Labudzynskyi, Shymanskyy, and Veliky 2016). Calbindin, whose expression was believed to be independent of hormones D, is a calcium binding protein that modulates intracellular calcium levels and influences the endometrial receptivity for implantation (Luu, Nie, and Salamonsen 2004, Nie et al. 2000). Endometrial expression of calbindin mRNA is indeed highly regulated during this phase. Osteopontin, also known as secreted phosphoprotein 1 (SPP1), participates in implantation and decidualization processes (Johnson et al. 2003), acting as signal transductor at the uterine-placental interface and regulating immune cell behavior and cytokine production. Between the luteal phase and the 12th week of pregnancy, hormones D induce the synthesis of molecules with beneficial effects (activin A, VEGF, CRH – previously indicated as CRF) (Woeckel et al. 2013, Wierzbicka et al. 2016, Song, Li, and An 2017) and inhibits the production of those with detrimental effects (estrogens, TNF-α and other inflammatory cytokines) (Song, Li, and An 2017, Halder SK 2010, Krishnan et al. 2010, Barrera et al. 2015).

Hormones D also modulate directly the biosynthesis of progesterone. In human cumulus granulosa cells, they significantly enhance the expression of 3β-HSD, which converts pregnenolone into progesterone, increasing its the local concentration (Merhi et al. 2014). Analogous behavior is observed in cultures of human ovarian cells (Parikh et al. 2010).

Progesterone

The receptor of progesterone is expressed in the glandular epithelium and the stroma of human endometrium during the menstrual cycle (Mote et al. 1999).

Progesterone and 1,25(OH)$_2$D mutually affect each other's metabolism. The former induces the expression of VDR gene in human T cells (Treg, Th1 and Th17, in various cytokine conditions; spleen CD4+) and in the uterus of pregnant or progestin-injected mice (Thangamani et al. 2015). Experiments using mifepristone (RU486), an antagonist of progesterone receptor, demonstrated that this effect is highly specific.

Progesterone-induced overexpression of VDR increases the sensitivity of T cells to 1,25(OH)$_2$D in different species, modulating the immune response that may prevent embryo implantation. 1,25(OH)$_2$D regulation of T cells, in fact, leads to the increase of Tregs and their associated molecules (FoxP3, CD8, LAP-TGF-β1), and to the suppression of effector T cells, which exert a pro-inflammatory activity. If siRNA is used to knock down the VDR gene, progesterone and 1,25(OH)$_2$D significantly lose the ability to inhibit T cell differentiation into effector T cells (Thangamani et al. 2015).

Progesterone also affects the response of the immune system, with mechanisms similar to hormones D (Figure 2): promotes the formation of Th2 cells and the release of their anti-inflammatory cytokines (Piccinni et al. 1995); regulates the differentiation of fetal T cells into Treg cells in human naive cord blood (Lee et al. 2011); suppresses the genes related to Th1 and Th17, concurrently enhancing Th2-related genes in lymphocytes from pregnant cows (Maeda et al. 2013), where it also inhibits the release of IFN-γ and IL-17 in a dose-dependent manner (Maeda et al. 2013); prevents degranulation and release of perforin from of NK cells, neutralizing their cytotoxicity (Druckmann and Druckmann 2005); induces the release of osteopontin in human decidua (Qu et al. 2008) and calbindin in the uterus of Egyptian buffalo (Emam, Abouelroos, and Gad 2016); between the luteal phase and the 12th week of pregnancy, stimulates the synthesis of molecules with beneficial effects (activin A, CRH, VEGF), while suppressing the release of those with detrimental effects (estrogens or TNF-α) (Druckmann and Druckmann 2005, Florio et al. 2003, Makrigiannakis et al. 1999, Kim et al. 2013, Kumar and Magon 2012). In mononuclear cells from human placental blood, progesterone antagonizes the effects of lipopolysaccharide (LPS) and significantly reduces the release of TNF-α, IL-1β, IL-6, IL-8, and

MIP-1α. In-vitro experiments showed increased levels of IL-10 after treatment with LPS and progesterone (Preciado-Martinez et al. 2018).

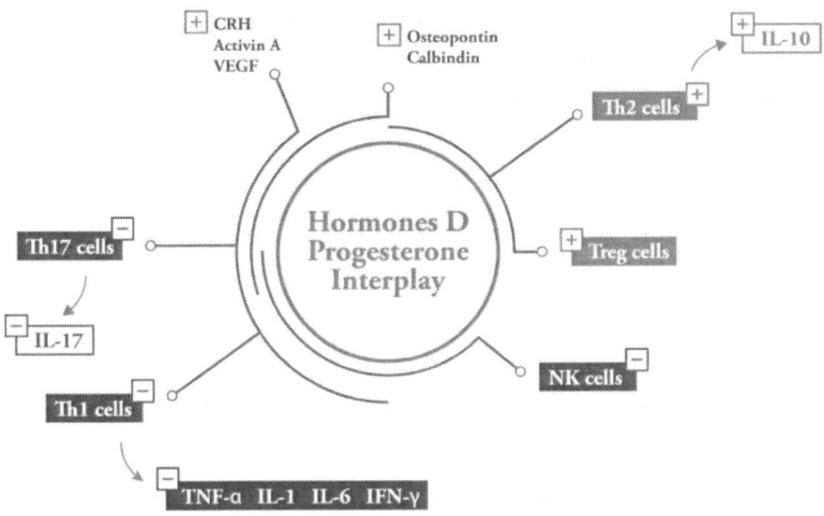

Figure 2. Common immunomodulatory activities exerted by hormones D and progesterone in early pregnancy. Hormones D and progesterone induce anti-inflammatory pathways (Th2 and Treg cells) and stimulate the release of corticotropin-releasing hormone (CRH), activin A, vascular endothelial growth factor (VEGF), osteopontin and calbindin. Concurrently, they inhibit pro-inflammatory pathways (NK, Th1 and Th17 cells), supporting pregnancy since its inception.

This evidence indicates that progesterone and hormones D act in a combined manner, mutually potentiating their anti-inflammatory activities to sustain early pregnancy. For a review on the topics discussed above, see the paper by Monastra et al. (Monastra et al. 2018).

Role in Advanced Pregnancy: The "Time Factor"

The role of hormones D and progesterone drastically changes through the course of pregnancy. While they increase the levels of corticotrophin-releasing hormone (CRH) in the initial phase, they inhibit its biosynthesis in advanced stages of gestation (Vrachnis et al. 2012, Wang et al. 2018), preventing preterm birth (PTB). CRH interaction with its receptor induces the transition from uterine quiescence to myometrial contractions and stimulates the fetal pituitary gland to produce adrenocorticotropic hormone (ACTH). ACTH increases the production of dehydroepiandrosterone sulphate (DHEA-S) and cortisol from the fetal adrenal glands. In the placenta, conversion of DHEA-S to estrogens causes myometrial contractions, while cortisol stimulates the production of surfactant protein A (SP-A) in fetal lungs, inducing uterine contractility and facilitating delivery.

Administration of "Vitamin" D in Pregnancy and in Assisted Reproductive Technology (ART)

Low maternal 25(OH)D status represents a risk factor for PTB (Qin et al. 2016), like both fetal and maternal VDR polymorphisms (Tolppanen et al. 2012). Physiologically, labor is a self-regulated inflammatory process, influenced by the balance between innate and adaptive immune cells (Bollapragada et al. 2009). Depleted circulating 25(OH)D concentration indicates a pro-inflammatory immune response, which increases the risk for PTB occurrence (Muller, Diamant, and Bendtzen 1991, Muller, Odum, and Bendtzen 1993).

Low hormones D levels in pregnancy correlate to reduced fetal growth and to adverse neonatal outcomes, in particular shortened gestational age (SGA) and diminished head circumference (HC). Moreover, maternal vitamin D deficiency (<30 nmol/L) exerts detrimental effects on offspring anthropometric parameters. Such condition is associated with lower birthweight and HC in their offspring (compared to mothers with vitamin D ≥30 nmol/L). While vitamin D deficiency and insufficiency (<50 nmol/L) bring about a higher risk of SGA and PTB, maternal 25(OH)D concentration ≥75 nmol/L is unrelated to altered birthweight, SGA and PTB (Tous et al. 2019).

Such outcomes are independent of ethnicity, but seem to be recurrent in pregnancies whose first and second trimester fall in winter or spring (Staples, Ponsonby, and Lim 2010). PTB risk, assessed against vitamin D supplementation, is significantly reduced in pregnant women with replenished values of 25(OH)D. In particular, subjects that reach 25(OH)D values $\geq$ 40 ng/ml prior to delivery have a 60% lower PTB occurrence than those with serum values < 40 ng/ml (McDonnell et al. 2017).

Hormones D are fundamental for successful pregnancy progression, and low levels are associated with risk of miscarriage and recurrent pregnancy loss. A study on 1072 women in childbearing age demonstrated a relationship between low hormone D status and infertility (Pagliardini et al. 2015). Analysis of 25(OH)D levels before treatment protocol for intracytoplasmic sperm injection (ICSI) revealed a positive significant correlation with number of oocytes, fertilized oocytes and endometrial thickness. In particular, high levels of 25(OH)D are necessary to develop the proper endometrial thickness for embryo implantation after ICSI. The correlation between 25(OH)D concentrations and clinical pregnancy remains a controversial topic. Indeed, while some authors state that clinical pregnancy rates increase with higher 25(OH)D levels, others report opposite effects (Ozkan et al. 2010, Anifandis et al. 2010, Firouzabadi et al. 2014, Ciepiela et al. 2018). Improved fertility and fetal conditions (Kwak-Kim et al. 2016), as well as successful ART results (Abdullah et al. 2017, Chu et al. 2018), were achieved restoring optimal hormone D status through supplementation (Hollis and Wagner 2017). Concerning the safety in pregnant women, vitamin D supplementation in doses up to 4,000 IU/day for 24-28 weeks proves to be completely safe (Hollis et al. 2011, Dawodu et al. 2013), without evidence of impact on calcium homeostasis or onset of bone pathologies.

CONCLUSION

Nowadays, 1,25$(OH)_2$D and 25(OH)D are considered hormones on the basis of several features that they share with steroid hormones, and the classification as vitamins may be considered outdated. In the case of insufficient exposure to sunlight, or in the presence of specific genetic variants, the endogenous synthesis of compounds D must be integrated with a dietary supply. Only in such cases the term "vitamin" is more appropriate. Compounds D belong to a group of molecules called secosteroids, which appeared early in the evolution of life to play a protective role for the living organism against UVB radiations and microbial infections. Later, the activities of these compounds evolved to maintain the physiological calcium levels, which are essential for several metabolic functions and for the formation of the skeleton. VDR and DBP, appeared and evolved separately, probably playing an independent protective role for the living forms in the early steps of evolution. They eventually converged at a certain point in the evolution timeline, becoming essential for the fully evolved hormone D activity.

Many preclinical researches and clinical studies demonstrated that hormones D play several physiological key roles in different areas, and that their essential function is not limited to proper bone growth and maintenance, as believed in the past. Their deficiency causes several detrimental effects on our health that can be addressed with "vitamin" D administration. The anti-inflammatory, antiproliferative and antibacterial activities of hormones D proved indeed very effective in the treatment of several systemic illnesses in the field of endocrinology, immunology, oncology, rheumatology, hematology, neurology, dermatology, and obstetrics/reproductive medicine.

$1,25(OH)_2D$ has a progesterone-like activity and plays essential roles in the female reproductive system, supporting the establishment and the proper development of pregnancy from the luteal phase of the menstrual cycle onward. Progesterone and $1,25(OH)_2D$ modulate the immune response associated with the inflammatory state in early pregnancy and prepare the endometrium receptivity, favoring embryo implantation. The two molecules regulate each other's metabolism and exert a combined action (Kim 2015), providing the basis for a successful pregnancy. In cases of low serum levels of hormones D, also in assisted reproductive procedures, vitamin D supplementation proved beneficial in restoring a physiological state. Replete hormone D status proved to reduce the risk of miscarriages and negative pregnancy outcomes.

REFERENCES

Abdullah, U. H., S. Lalani, F. Syed, S. Arif, and R. Rehman. 2017. "Association of Vitamin D with outcome after intra cytoplasmic sperm injection." *J Matern Fetal Neonatal Med* 30 (1):117-120. doi: 10.3109/14767058.2016.1163680.

Acker, G. M., M. Garabedian, H. Guillozo, M. A. Pecquinot, and S. Balsan. 1982. "In vitro formation of 25-hydroxyvitamin D3 metabolites in endometrium: dependence on the hormonal status of the rat." *Endocrinology* 111 (6):2103-9. doi: 10.1210/endo-111-6-2103.

Anifandis, G. M., K. Dafopoulos, C. I. Messini, N. Chalvatzas, N. Liakos, S. Pournaras, and I. E. Messinis. 2010. "Prognostic value of follicular fluid 25-OH vitamin D and glucose levels in the IVF outcome." *Reprod Biol Endocrinol* 8:91. doi: 10.1186/1477-7827-8-91.

Barrera, D., L. Diaz, N. Noyola-Martinez, and A. Halhali. 2015. "Vitamin D and Inflammatory Cytokines in Healthy and Preeclamptic Pregnancies." *Nutrients* 7 (8):6465-90. doi: 10.3390/nu7085293.

Berridge, M. J. 2017. "Vitamin D deficiency accelerates ageing and age-related diseases: a novel hypothesis." *J Physiol* 595 (22):6825-6836. doi: 10.1113/jp274887.

Bikle, D. D. 2011. "Vitamin D: an ancient hormone." *Exp Dermatol* 20 (1):7-13.

Bikle, D. D., S. Malmstroem, and J. Schwartz. 2017. "Current Controversies: Are Free Vitamin Metabolite Levels a More Accurate Assessment of Vitamin D Status than Total Levels?" *Endocrinol Metab Clin North Am* 46 (4):901-918. doi: 10.1016/j.ecl.2017.07.013.

Bikle, D. D., and J. Schwartz. 2019. "Vitamin D Binding Protein, Total and Free Vitamin D Levels in Different Physiological and Pathophysiological Conditions." *Front Endocrinol (Lausanne)* 10:317.doi:10.3389/fendo.2019.00317.

Blomberg Jensen, M. 2014. "Vitamin D and male reproduction." *Nat Rev Endocrinol* 10 (3):175-86. doi: 10.1038/nrendo.2013.262.

Boisen, I. M., L. Bollehuus Hansen, L. J. Mortensen, B. Lanske, A. Juul, and M. Blomberg Jensen. 2017. "Possible influence of vitamin D on male reproduction." *J Steroid Biochem Mol Biol* 173:215-222. doi:10.1016/j.jsbmb.2016.09.023.

Bollapragada, S., R. Youssef, F. Jordan, I. Greer, J. Norman, and S. Nelson. 2009. "Term labor is associated with a core inflammatory response in human fetal membranes, myometrium, and cervix." *Am J Obstet Gynecol* 200 (1):104.e1-11. doi: 10.1016/j.ajog.2008.08.032.

Borel, P., D. Caillaud, and N. J. Cano. 2015. "Vitamin D bioavailability: state of the art." *Crit Rev Food Sci Nutr* 55 (9):1193-205. doi: 10.1080/10408398.2012.688897.

Bouillon, R., and T. Suda. 2014. "Vitamin D: calcium and bone homeostasis during evolution." *Bonekey Rep* 3:480. doi: 10.1038/bonekey.2013.214.

Brot, C., L. B. Jensen, and O. H. Sorensen. 1997. "Bone mass and risk factors for bone loss in perimenopausal Danish women." *J Intern Med* 242 (6):505-11. doi: 10.1111/j.1365-2796.1997.tb00024.x.

Brown, A. J., and A. M. Galea. 2010. "Cholesterol as an evolutionary response to living with oxygen." *Evolution* 64 (7):2179-83. doi:10.1111/j.1558-5646.2010.01011.x.

Carlberg, C., S. Seuter, and S. Heikkinen. 2012. "The first genome-wide view of vitamin D receptor locations and their mechanistic implications." *Anticancer Res* 32 (1):271-82.

Carpenter, K. J., and L. Zhao. 1999. "Forgotten mysteries in the early history of vitamin D." *J Nutr* 129 (5):923-7. doi: 10.1093/jn/129.5.923.

Chakraborty, I., S. K. Das, and S. K. Dey. 1995. "Differential expression of vascular endothelial growth factor and its receptor mRNAs in the mouse uterus around the time of implantation." *J Endocrinol* 147 (2):339-52.

Chambon, Y. 1951. "[Synergic action of vitamin D and of progesterone in obtaining ovum implantation in the castrated pregnant rabbit]." *C R Seances Soc Biol Fil* 145 (13-14):955-9.

Chishimba, L., D. R. Thickett, R. A. Stockley, and A. M. Wood. 2010. "The vitamin D axis in the lung: a key role for vitamin D-binding protein." *Thorax* 65 (5):456-62. doi: 10.1136/thx.2009.128793.

Choi, K. C., P. C. Leung, and E. B. Jeung. 2005. "Biology and physiology of Calbindin-D9k in female reproductive tissues: involvement of steroids and endocrine disruptors." *Reprod Biol Endocrinol* 3:66. doi: 10.1186/1477-7827-3-66.

Choudhary, D., I. Jansson, I. Stoilov, M. Sarfarazi, and J. B. Schenkman. 2005. "Expression patterns of mouse and human CYP orthologs (families 1-4) during development and in different adult tissues." *Arch Biochem Biophys* 436 (1):50-61. doi: 10.1016/j.abb.2005.02.001.

Christakos, S., M. Hewison, D. G. Gardner, C. L. Wagner, I. N. Sergeev, E. Rutten, A. G. Pittas, R. Boland, L. Ferrucci, and D. D. Bikle. 2013. "Vitamin D: beyond bone." *Ann N Y Acad Sci* 1287:45-58. doi:10.1111/nyas.12129.

Christodoulou, S., T. Goula, A. Ververidis, and G. Drosos. 2013. "Vitamin D and bone disease." *Biomed Res Int* 2013:396541. doi: 10.1155/2013/396541.

Chu, J., I. Gallos, A. Tobias, B. Tan, A. Eapen, and A. Coomarasamy. 2018. "Vitamin D and assisted reproductive treatment outcome: a systematic review and meta-analysis." *Hum Reprod* 33 (1):65-80. doi:10.1093/humrep/dex326.

Ciepiela, P., A. J. Duleba, E. Kowaleczko, K. Chelstowski, and R. Kurzawa. 2018. "Vitamin D as a follicular marker of human oocyte quality and a serum marker of in vitro fertilization outcome." *J Assist Reprod Genet* 35 (7):1265-1276. doi: 10.1007/s10815-018-1179-4.

Cook J. W., Dodds E. C., Hewett C. L., Lawson W. 1934. "The Oestrogenic Activity of some Condensed-Ring Compounds in Relation to their other Biological Activities." *Proceedings of the Royal Society of London. Series B* 114:272-286.

Cooke, N. E., and J. G. Haddad. 1989. "Vitamin D binding protein (Gc-globulin)." *Endocr Rev* 10 (3):294-307. doi: 10.1210/edrv-10-3-294.

Darwish, H., and H. F. DeLuca. 1993. "Vitamin D-regulated gene expression." *Crit Rev Eukaryot Gene Expr* 3 (2):89-116.

Davis, J. S., and B. R. Rueda. 2002. "The corpus luteum: an ovarian structure with maternal instincts and suicidal tendencies." *Front Biosci* 7:d1949-78.

Dawodu, A., H. F. Saadi, G. Bekdache, Y. Javed, M. Altaye, and B. W. Hollis. 2013. "Randomized controlled trial (RCT) of vitamin D supplementation in pregnancy in a population with endemic vitamin D deficiency." *J Clin Endocrinol Metab* 98 (6):2337-46. doi: 10.1210/jc. 2013-1154.

Delanghe, J. R., R. Speeckaert, and M. M. Speeckaert. 2015. "Behind the scenes of vitamin D binding protein: more than vitamin D binding." *Best Pract Res Clin Endocrinol Metab* 29 (5):773-86. doi: 10.1016/j.beem.2 015.06.006.

DeLuca, H. F. 2004. "Overview of general physiologic features and functions of vitamin D." *Am J Clin Nutr* 80 (6 Suppl):1689S-96S. doi: 10.1093/ajcn/80.6.1689S.

Delvin, E. E., A. Arabian, F. H. Glorieux, and O. A. Mamer. 1985. "In vitro metabolism of 25-hydroxycholecalciferol by isolated cells from human decidua." *J Clin Endocrinol Metab* 60 (5):880-5. doi: 10.1210/jcem-60-5-880.

Di Somma, C., E. Scarano, L. Barrea, V. V. Zhukouskaya, S. Savastano, C. Mele, M. Scacchi, G. Aimaretti, A. Colao, and P. Marzullo. 2017. "Vitamin D and Neurological Diseases: An Endocrine View." *Int J Mol Sci* 18 (11). doi: 10.3390/ijms18112482.

Druckmann, R., and M. A. Druckmann. 2005. "Progesterone and the immunology of pregnancy." *J Steroid Biochem Mol Biol* 97 (5):389-96. doi: 10.1016/j.jsbmb.2005.08.010.

Dunn, P. M. 1998. "Francis Glisson (1597-1677) and the "discovery" of rickets." *Arch Dis Child Fetal Neonatal Ed* 78 (2):F154-5. doi:10.113 6/fn.78.2.f154.

Emam, M. A., M. E. Abouelroos, and F. A. Gad. 2016. "Expression of calbindin-D9k and vitamin D receptor in the uterus of Egyptian buffalo during follicular and luteal phases." *Acta Histochem* 118 (5):471-7. doi: 10.1016/j.acthis.2016.04.009.

Evans, K. N., L. Nguyen, J. Chan, B. A. Innes, J. N. Bulmer, M. D. Kilby, and M. Hewison. 2006. "Effects of 25-hydroxyvitamin D3 and 1,25-dihydroxyvitamin D3 on cytokine production by human decidual cells." *Biol Reprod* 75 (6):816-22. doi: 10.1095/biolreprod.106.054056.

Fawaz, L., M. F. Mrad, J. M. Kazan, S. Sayegh, R. Akika, and S. J. Khoury. 2016. "Comparative effect of 25(OH)D3 and 1,25(OH)2D3 on Th17 cell differentiation." *Clin Immunol* 166-167:59-71. doi: 10.1016/j.clim.2016.02.011.

Feldman, D., A. V. Krishnan, S. Swami, E. Giovannucci, and B. J. Feldman. 2014. "The role of vitamin D in reducing cancer risk and progression." *Nat Rev Cancer* 14 (5):342-57. doi: 10.1038/nrc3691.

Feyel-Cabanes, T. 1949. "Maintien de la grossesse chez le lapine gestante castrée par une association hormone-stérol." *C R Seances Soc Biol Fil.* 143:1454-1458.

Firouzabadi, R. D., E. Rahmani, M. Rahsepar, and M. M. Firouzabadi. 2014. "Value of follicular fluid vitamin D in predicting the pregnancy rate in an IVF program." *Arch Gynecol Obstet* 289 (1):201-6. doi:10.1007/s00404-013-2959-9.

Florio, P., M. Rossi, M. Sigurdardottir, P. Ciarmela, S. Luisi, P. Vigano, D. Grasso, G. Fiore, L. Cobellis, A. M. Di Blasio, and F. Petraglia. 2003. "Paracrine regulation of endometrial function: interaction between progesterone and corticotropin-releasing factor (CRF) and activin A." *Steroids* 68 (10-13):801-7.

Foresta, C., R. Selice, A. Di Mambro, and G. Strapazzon. 2010. "Testiculopathy and vitamin D insufficiency." *Lancet* 376 (9749):1301. doi: 10.1016/s0140-6736(10)61916-2.

Forrest, K. Y., and W. L. Stuhldreher. 2011. "Prevalence and correlates of vitamin D deficiency in US adults." *Nutr Res* 31 (1):48-54. doi: 10.1016/j.nutres.2010.12.001.

Goring, H., and S. Koshuchowa. 2015. "Vitamin D -- the sun hormone. Life in environmental mismatch." *Biochemistry (Mosc)* 80 (1):8-20. doi: 10.1134/S0006297915010022.

Gorman, S., S. Geldenhuys, M. Judge, C. E. Weeden, J. Waithman, and P. H. Hart. 2016. "Dietary Vitamin D Increases Percentages and Function of Regulatory T Cells in the Skin-Draining Lymph Nodes and Suppresses Dermal Inflammation." *J Immunol Res* 2016:1426503. doi: 10.1155/2016/1426503.

Guy, R. A. 1923. "The history of cod liver oil as a remedy." *Am J Dis Child.* 26:112–116.

Haddad, J. G., Jr., and S. Rojanasathit. 1976. "Acute administration of 25-hydroxycholecalciferol in man." *J Clin Endocrinol Metab* 42 (2):284-90. doi: 10.1210/jcem-42-2-284.

Halder SK, Goodwin S, Al-Hendy A. 2010. "Vitamin D exhibits antiestrogenic effects in human uterine leiomyoma cells." *Fertil Steril* 94:S219–S220.

Harrison, S. R., D. Li, L. E. Jeffery, K. Raza, and M. Hewison. 2019. "Vitamin D, Autoimmune Disease and Rheumatoid Arthritis." *Calcif Tissue Int*. doi: 10.1007/s00223-019-00577-2.

Hartwell, D., B. J. Riis, and C. Christiansen. 1990. "Changes in vitamin D metabolism during natural and medical menopause." *J Clin Endocrinol Metab* 71 (1):127-32. doi: 10.1210/jcem-71-1-127.

Haussler, M. R. 1986. "Vitamin D receptors: nature and function." *Annu Rev Nutr* 6:527-62. doi: 10.1146/annurev.nu.06.070186.002523.

Haussler, M. R., P. W. Jurutka, M. Mizwicki, and A. W. Norman. 2011. "Vitamin D receptor (VDR)-mediated actions of 1alpha, 25(OH)(2) vitamin D(3): genomic and non-genomic mechanisms." *Best Pract Res Clin Endocrinol Metab* 25 (4):543-59. doi:10.1016/j.beem.2011.05.010.

Haussler M. R., Myrtle J. F., Norman A. W. 1968. "The association of a metabolite of vitamin D3 with intestinal mucosa chromatin in vivo." *J Biol Chem* 243:4055–4064.

Haussler M. R., Norman A. W. 1969. "Chromosomal receptor for a vitamin D metabolite." *Proc Natl Acad Sci USA* 62:155–162.

Heaney, R. P., L. A. Armas, J. R. Shary, N. H. Bell, N. Binkley, and B. W. Hollis. 2008. "25-Hydroxylation of vitamin D3: relation to circulating vitamin D3 under various input conditions." *Am J Clin Nutr* 87 (6):1738-42. doi: 10.1093/ajcn/87.6.1738.

Hernigou, P., J. C. Auregan, and A. Dubory. 2018. "Vitamin D: part I; from plankton and calcified skeletons (500 million years ago) to rickets." *Int Orthop* 42 (9):2273-2285. doi: 10.1007/s00264-018-3857-3.

Herr, D., I. Bekes, and C. Wulff. 2013. "Regulation of Endothelial Permeability in the Corpus Luteum: A Review of the Literature." *Geburtshilfe Frauenheilkd* 73 (11):1107-1111. doi: 10.1055/s-0033-1351032.

Hewison, M., D. Zehnder, R. Bland, and P. M. Stewart. 2000. "1alpha-Hydroxylase and the action of vitamin D." *J Mol Endocrinol* 25 (2):141-8.

Holick, M. F., Schnoes H. K., DeLuca H. F. 1971. "Identification of 1,25-dihydroxycholecalciferol, a form of vitamin D3 metabolically active in the intestine." *Proc Natl Acad Sci U S A*. 68(4):803-4.

Holick, M. F. 2003. "Vitamin D: A millenium perspective." *J Cell Biochem* 88 (2):296-307. doi: 10.1002/jcb.10338.

Holick, M. F. 2011. "Vitamin D: evolutionary, physiological and health perspectives." *Curr Drug Targets* 12 (1):4-18. doi: 10.2174/138945 011793591635.

Holick, M. F. 2017. "The vitamin D deficiency pandemic: Approaches for diagnosis, treatment and prevention." *Rev Endocr Metab Disord* 18 (2):153-165. doi: 10.1007/s11154-017-9424-1.

Holick, M. F., T. C. Chen, Z. Lu, and E. Sauter. 2007. "Vitamin D and skin physiology: a D-lightful story." *J Bone Miner Res* 22 Suppl 2:V28-33. doi: 10.1359/jbmr.07s211.

Holick, M. F., X. Q. Tian, and M. Allen. 1995. "Evolutionary importance for the membrane enhancement of the production of vitamin D3 in the skin of poikilothermic animals." *Proc Natl Acad Sci U S A* 92 (8):3124-6. doi: 10.1073/pnas.92.8.3124.

Hollis, B. W., D. Johnson, T. C. Hulsey, M. Ebeling, and C. L. Wagner. 2011. "Vitamin D supplementation during pregnancy: double-blind, randomized clinical trial of safety and effectiveness." *J Bone Miner Res* 26 (10):2341-57. doi: 10.1002/jbmr.463.

Hollis, B. W., and C. L. Wagner. 2017. "Vitamin D supplementation during pregnancy: Improvements in birth outcomes and complications through direct genomic alteration." *Mol Cell Endocrinol* 453:113-130. doi: 10.1016/j.mce.2017.01.039.

Intakes, Standing Committee on the Scientific Evaluation of Dietary Reference. 1997. *Dietary reference intakes: calcium, phosphorus, magnesium, Vitamin D, and fluoride*. Washington (DC): National Academy Press.

Johnson, G. A., R. C. Burghardt, F. W. Bazer, and T. E. Spencer. 2003. "Osteopontin: roles in implantation and placentation." *Biol Reprod* 69 (5):1458-71. doi: 10.1095/biolreprod.103.020651.

Jones, G. 2008. "Pharmacokinetics of vitamin D toxicity." *Am J Clin Nutr* 88 (2):582s-586s. doi: 10.1093/ajcn/88.2.582S.

Jones, G., D. E. Prosser, and M. Kaufmann. 2012. "25-Hydroxyvitamin D-24-hydroxylase (CYP24A1): its important role in the degradation of vitamin D." *Arch Biochem Biophys* 523 (1):9-18.doi:10.1016/j.abb.2011.11.003.

Kachkache, M., C. Rebut-Bonneton, J. Demignon, E. Cynober, and M. Garabedian. 1993. "Uterine cells other than stromal decidual cells are required for 1,25-dihydroxyvitamin D3 production during early human pregnancy." *FEBS Lett* 333 (1-2):83-8.

Kim, C. H. 2015. "A functional relay from progesterone to vitamin D in the immune system." *DNA Cell Biol* 34 (6):379-82. doi: 10.1089/dna.2015.2857.

Kim, M., H. J. Park, J. W. Seol, J. Y. Jang, Y. S. Cho, K. R. Kim, Y. Choi, J. P. Lydon, F. J. Demayo, M. Shibuya, N. Ferrara, H. K. Sung, A. Nagy, K. Alitalo, and G. Y. Koh. 2013. "VEGF-A regulated by progesterone governs uterine angiogenesis and vascular remodelling during pregnancy." *EMBO Mol Med* 5 (9):1415-30. doi: 10.1002/emmm.201302618.

Krishnan, A. V., and D. Feldman. 2011. "Mechanisms of the anti-cancer and anti-inflammatory actions of vitamin D." *Annu Rev Pharmacol Toxicol* 51:311-36. doi: 10.1146/annurev-pharmtox-010510-100611.

Krishnan, A. V., S. Swami, L. Peng, J. Wang, J. Moreno, and D. Feldman. 2010. "Tissue-selective regulation of aromatase expression by calcitriol:

implications for breast cancer therapy." *Endocrinology* 151 (1):32-42. doi: 10.1210/en.2009-0855.

Kulling, P. M., K. C. Olson, T. L. Olson, D. J. Feith, and T. P. Loughran, Jr. 2017. "Vitamin D in hematological disorders and malignancies." *Eur J Haematol* 98 (3):187-197. doi: 10.1111/ejh.12818.

Kumar, P., and N. Magon. 2012. "Hormones in pregnancy." *Niger Med J* 53 (4):179-83. doi: 10.4103/0300-1652.107549.

Kumar, R. 1980. "The metabolism of 1,25-dihydroxyvitamin D3." *Endocr Rev* 1 (3):258-67. doi: 10.1210/edrv-1-3-258.

Kwak-Kim, J., A. Skariah, L. Wu, D. Salazar, N. Sung, and K. Ota. 2016. "Humoral and cellular autoimmunity in women with recurrent pregnancy losses and repeated implantation failures: A possible role of vitamin D." *Autoimmun Rev* 15 (10):943-7. doi: 10.1016/j.autrev.2016.07.015.

Labudzynskyi, D., I. Shymanskyy, and M. Veliky. 2016. "Role of vitamin D3 in regulation of interleukin-6 and osteopontin expression in liver of diabetic mice." *Eur Rev Med Pharmacol Sci* 20 (13):2916-9.

Larque, E., E. Morales, R. Leis, and J. E. Blanco-Carnero. 2018. "Maternal and Foetal Health Implications of Vitamin D Status during Pregnancy." *Ann Nutr Metab* 72 (3):179-192. doi: 10.1159/000487370.

Lee, J. H., B. Ulrich, J. Cho, J. Park, and C. H. Kim. 2011. "Progesterone promotes differentiation of human cord blood fetal T cells into T regulatory cells but suppresses their differentiation into Th17 cells." *J Immunol* 187 (4):1778-87. doi: 10.4049/jimmunol.1003919.

Levin, E. R. 2015. "Extranuclear steroid receptors are essential for steroid hormone actions." *Annu Rev Med* 66:271-80. doi: 10.1146/annurev-med-050913-021703.

Levine, B. S., F. R. Singer, G. F. Bryce, J. P. Mallon, O. N. Miller, and J. W. Coburn. 1985. "Pharmacokinetics and biologic effects of calcitriol in normal humans." *J Lab Clin Med* 105 (2):239-46.

Lips, P. 2006. "Vitamin D physiology." *Prog Biophys Mol Biol* 92 (1):4-8. doi: 10.1016/j.pbiomolbio.2006.02.016.

Liu, W., L. Zhang, H. J. Xu, Y. Li, C. M. Hu, J. Y. Yang, and M. Y. Sun. 2018. "The Anti-Inflammatory Effects of Vitamin D in Tumorigenesis." *Int J Mol Sci* 19 (9). doi: 10.3390/ijms19092736.

Lorenzen, M., I. M. Boisen, L. J. Mortensen, B. Lanske, A. Juul, and M. Blomberg Jensen. 2017. "Reproductive endocrinology of vitamin D." *Mol Cell Endocrinol* 453:103-112. doi: 10.1016/j.mce.2017.03.023.

Luu, K. C., G. Y. Nie, and L. A. Salamonsen. 2004. "Endometrial calbindins are critical for embryo implantation: evidence from in vivo use of morpholino antisense oligonucleotides." *Proc Natl Acad Sci U S A* 101 (21):8028-33. doi: 10.1073/pnas.0401069101.

Maeda, Y., H. Ohtsuka, M. Tomioka, and M. Oikawa. 2013. "Effect of progesterone on Th1/Th2/Th17 and regulatory T cell-related genes in peripheral blood mononuclear cells during pregnancy in cows." *Vet Res Commun* 37 (1):43-9. doi: 10.1007/s11259-012-9545-7.

Makrigiannakis, A., A. N. Margioris, E. Chatzaki, E. Zoumakis, G. P. Chrousos, and A. Gravanis. 1999. "The decidualizing effect of progesterone may involve direct transcriptional activation of corticotrophin-releasing hormone from human endometrial stromal cells." *Mol Hum Reprod* 5 (9):789-96.

Marino, R., and M. Misra. 2019. "Extra-Skeletal Effects of Vitamin D." *Nutrients* 11 (7). doi: 10.3390/nu11071460.

Mawer, E. B., J. Backhouse, C. A. Holman, G. A. Lumb, and S. W. Stanbury. 1972. "The distribution and storage of vitamin D and its metabolites in human tissues." *Clin Sci* 43 (3):413-31.

McCollum, E. V.; Simmonds, N.; Becker, J. E.; Shipley, P. G. 1922. "Studies on the experimental rickets; and experimental demonstration of the existence of a vitamin which promotes calcium deposition" *J Biol Chem* 53:293–312.

McDonnell, S. L., K. A. Baggerly, C. A. Baggerly, J. L. Aliano, C. B. French, L. L. Baggerly, M. D. Ebeling, C. S. Rittenberg, C. G. Goodier, J. F. Mateus Nino, R. J. Wineland, R. B. Newman, B. W. Hollis, and C. L. Wagner. 2017. "Maternal 25(OH)D concentrations >/=40 ng/mL associated with 60% lower preterm birth risk among general obstetrical patients at an urban medical center." *PLoS One* 12 (7):e0180483. doi: 10.1371/journal.pone.0180483.

Merhi, Z., A. Doswell, K. Krebs, and M. Cipolla. 2014. "Vitamin D alters genes involved in follicular development and steroidogenesis in human cumulus granulosa cells." *J Clin Endocrinol Metab* 99 (6):E1137-45. doi: 10.1210/jc.2013-4161.

Mesen, T. B., and S. L. Young. 2015. "Progesterone and the luteal phase: a requisite to reproduction." *Obstet Gynecol Clin North Am* 42 (1):135-51. doi: 10.1016/j.ogc.2014.10.003.

Mizwicki, M. T., D. Menegaz, S. Yaghmaei, H. L. Henry, and A. W. Norman. 2010. "A molecular description of ligand binding to the two overlapping binding pockets of the nuclear vitamin D receptor (VDR): structure-function implications." *J Steroid Biochem Mol Biol* 121 (1-2):98-105. doi: 10.1016/j.jsbmb.2010.04.005.

Monastra, G., S. De Grazia, L. De Luca, S. Vittorio, and V. Unfer. 2018. "Vitamin D: a steroid hormone with progesterone-like activity." *Eur Rev Med Pharmacol Sci* 22 (8):2502-2512. doi: 10.26355/eurrev_201804_14845.

Mote, P. A., R. L. Balleine, E. M. McGowan, and C. L. Clarke. 1999. "Colocalization of progesterone receptors A and B by dual immunofluorescent histochemistry in human endometrium during the menstrual cycle." *J Clin Endocrinol Metab* 84 (8):2963-71. doi: 10.1210/jcem.84.8.5928.

Moussa, M., J. F. Landrier, E. Reboul, O. Ghiringhelli, C. Comera, X. Collet, K. Frohlich, V. Bohm, and P. Borel. 2008. "Lycopene absorption in human intestinal cells and in mice involves scavenger receptor class B type I but not Niemann-Pick C1-like 1." *J Nutr* 138 (8):1432-6. doi: 10.1093/jn/138.8.1432.

Muller, K., M. Diamant, and K. Bendtzen. 1991. "Inhibition of production and function of interleukin-6 by 1,25-dihydroxyvitamin D3." *Immunol Lett* 28 (2):115-20. doi: 10.1016/0165-2478(91)90108-m.

Muller, K., N. Odum, and K. Bendtzen. 1993. "1,25-dihydroxyvitamin D3 selectively reduces interleukin-2 levels and proliferation of human T cell lines in vitro." *Immunol Lett* 35 (2):177-82. doi: 10.1016/0165-2478(93)90088-j.

Myrtle JF, Haussler MR, Norman AW. . 1970. "Evidence for the biologically active form of cholecalciferol in the intestine." *J Biol Chem* 245:1190–1196.

Narushima, K., T. Takada, Y. Yamanashi, and H. Suzuki. 2008. "Niemann-pick C1-like 1 mediates alpha-tocopherol transport." *Mol Pharmacol* 74 (1):42-9. doi: 10.1124/mol.107.043034.

Newmark, H., W. Dantoft, and P. Ghazal. 2017. "Evolutionary Origin of the Interferon-Immune Metabolic Axis: The Sterol-Vitamin D Link." *Front Immunol* 8:62. doi: 10.3389/fimmu.2017.00062.

Nezbedova, P., and J. Brtko. 2004. "1alpha,25-dihydroxyvitamin D3 inducible transcription factor and its role in the vitamin D action." *Endocr Regul* 38 (1):29-38.

Nie, G. Y., Y. Li, J. Wang, H. Minoura, J. K. Findlay, and L. A. Salamonsen. 2000. "Complex regulation of calcium-binding protein D9k (calbindin-D(9k)) in the mouse uterus during early pregnancy and at the site of embryo implantation." *Biol Reprod* 62 (1):27-36. doi: 10.1095/biolreprod62.1.27.

Niswender, G. D., J. L. Juengel, P. J. Silva, M. K. Rollyson, and E. W. McIntush. 2000. "Mechanisms controlling the function and life span of the corpus luteum." *Physiol Rev* 80 (1):1-29. doi: 10.1152/physrev.2000.80.1.1.

Norman, A. W. 2006. "Minireview: vitamin D receptor: new assignments for an already busy receptor." *Endocrinology* 147 (12):5542-8. doi: 10.1210/en.2006-0946.

Norman, A. W. 2012. "The history of the discovery of vitamin D and its daughter steroid hormone." *Ann Nutr Metab* 61 (3):199-206. doi: 10.1159/000343104.

Norman, A. W., M. T. Mizwicki, and D. P. Norman. 2004. "Steroid-hormone rapid actions, membrane receptors and a conformational ensemble model." *Nat Rev Drug Discov* 3 (1):27-41. doi: 10.1038/nrd1283.

Norman, A. W.; Myrtle, J. F.; Midgett, R. J.; Nowicki, H. G.; Williams, V.; Popjak, G. 1971. "1,25-dihydroxycholecalciferol: identification of the proposed active form of vitamin D3 in the intestine." *Science* 17351-54:51-54.

O'Malley, B. W. 1971. "Mechanisms of action of steroid hormones." *N Engl J Med* 284 (7):370-7. doi: 10.1056/NEJM197102182840710.

O'Riordan, J. L., and O. L. Bijvoet. 2014. "Rickets before the discovery of vitamin D." *Bonekey Rep* 3:478. doi: 10.1038/bonekey.2013.212.

Ota, K., S. Dambaeva, M. W. Kim, A. R. Han, A. Fukui, A. Gilman-Sachs, K. Beaman, and J. Kwak-Kim. 2015. "1,25-Dihydroxy-vitamin D3 regulates NK-cell cytotoxicity, cytokine secretion, and degranulation in women with recurrent pregnancy losses." *Eur J Immunol* 45 (11):3188-99. doi: 10.1002/eji.201545541.

Ozkan, S., S. Jindal, K. Greenseid, J. Shu, G. Zeitlian, C. Hickmon, and L. Pal. 2010. "Replete vitamin D stores predict reproductive success following in vitro fertilization." *Fertil Steril* 94 (4):1314-9. doi: 10.1016/j.fertnstert.2009.05.019.

Pagliardini, L., P. Vigano, M. Molgora, P. Persico, A. Salonia, S. H. Vailati, A. Paffoni, E. Somigliana, E. Papaleo, and M. Candiani. 2015. "High Prevalence of Vitamin D Deficiency in Infertile Women Referring for Assisted Reproduction." *Nutrients* 7 (12):9972-84. doi: 10.3390/nu7125516.

Parikh, G., M. Varadinova, P. Suwandhi, T. Araki, Z. Rosenwaks, L. Poretsky, and D. Seto-Young. 2010. "Vitamin D regulates steroidogenesis and insulin-like growth factor binding protein-1 (IGFBP-1) production in human ovarian cells." *Horm Metab Res* 42 (10):754-7. doi: 10.1055/s-0030-1262837.

Petkovich M., Helvig C., Epps T. 2011. *CYP24A1 Regulation in Health and Disease*. Edited by Pike JW Feldman D, Adams JS. Third Edition ed. Vol. 2, *Vitamin D*. San Diego: Academic Press.

Piazzolla, G., A. Castrovilli, V. Liotino, M. R. Vulpi, M. Fanelli, A. Mazzocca, M. Candigliota, E. Berardi, O. Resta, C. Sabba, and C. Tortorella. 2017. "Metabolic syndrome and Chronic Obstructive Pulmonary Disease (COPD): The interplay among smoking, insulin resistance and vitamin D." *PLoS One* 12 (10):e0186708. doi: 10.1371/journal.pone.0186708.

Piccinni, M. P., M. G. Giudizi, R. Biagiotti, L. Beloni, L. Giannarini, S. Sampognaro, P. Parronchi, R. Manetti, F. Annunziato, C. Livi, and et al. 1995. "Progesterone favors the development of human T helper cells producing Th2-type cytokines and promotes both IL-4 production and membrane CD30 expression in established Th1 cell clones." *J Immunol* 155 (1):128-33.

Poon, A. H., L. Gong, C. Brasch-Andersen, A. A. Litonjua, B. A. Raby, Q. Hamid, C. Laprise, S. T. Weiss, R. B. Altman, and T. E. Klein. 2012. "Very important pharmacogene summary for VDR." *Pharmacogenet Genomics* 22 (10):758-63. doi: 10.1097/FPC.0b013e328354455c.

Preciado-Martinez, E., G. Garcia-Ruiz, P. Flores-Espinosa, L. Bermejo-Martinez, A. Espejel-Nunez, G. Estrada-Gutierrez, G. Razo-Aguilera, M. Granados-Cepeda, A. C. Helguera-Repetto, C. Irles, and V. Zaga-Clavellina. 2018. "Progesterone suppresses the lipopolysaccharide-induced pro-inflammatory response in primary mononuclear cells isolated from human placental blood." *Immunol Invest* 47 (2):181-195. doi: 10.1080/08820139.2017.1413112.

Prosser, D. E., and G. Jones. 2004. "Enzymes involved in the activation and inactivation of vitamin D." *Trends Biochem Sci* 29 (12):664-73. doi: 10.1016/j.tibs.2004.10.005.

Qin, L. L., F. G. Lu, S. H. Yang, H. L. Xu, and B. A. Luo. 2016. "Does Maternal Vitamin D Deficiency Increase the Risk of Preterm Birth: A Meta-Analysis of Observational Studies." *Nutrients* 8 (5). doi: 10.3390/nu8050301.

Qu, X., M. Yang, W. Zhang, L. Liang, Y. Yang, Y. Zhang, B. Deng, W. Gao, J. Liu, Q. Yang, B. Kong, and F. Gong. 2008. "Osteopontin expression in human decidua is associated with decidual natural killer cells recruitment and regulated by progesterone." *In Vivo* 22 (1):55-61.

Rafiee, M., M. Gharagozloo, A. Ghahiri, F. Mehrabian, M. R. Maracy, S. Kouhpayeh, I. L. Pieper, and A. Rezaei. 2015. "Altered Th17/Treg Ratio in Recurrent Miscarriage after Treatment with Paternal Lymphocytes and Vitamin D3: a Double-Blind Placebo-Controlled Study." *Iran J Immunol* 12 (4):252-62. doi: IJIv12i4A3.

Rajakumar, K. 2003. "Vitamin D, cod-liver oil, sunlight, and rickets: a historical perspective." *Pediatrics* 112 (2):e132-5. doi: 10.1542/pe ds.11 2.2.e132.

Reboul, E., L. Abou, C. Mikail, O. Ghiringhelli, M. Andre, H. Portugal, D. Jourdheuil-Rahmani, M. J. Amiot, D. Lairon, and P. Borel. 2005. "Lutein transport by Caco-2 TC-7 cells occurs partly by a facilitated process involving the scavenger receptor class B type I (SR-BI)." *Biochem J* 387 (Pt 2):455-61. doi: 10.1042/bj20040554.

Reboul, E., A. Klein, F. Bietrix, B. Gleize, C. Malezet-Desmoulins, M. Schneider, A. Margotat, L. Lagrost, X. Collet, and P. Borel. 2006. "Scavenger receptor class B type I (SR-BI) is involved in vitamin E transport across the enterocyte." *J Biol Chem* 281 (8):4739-45. doi: 10.1074/jbc.M509042200.

Robien, K., S. J. Oppeneer, J. A. Kelly, and J. M. Hamilton-Reeves. 2013. "Drug-vitamin D interactions: a systematic review of the literature." *Nutr Clin Pract* 28 (2):194-208. doi: 10.1177/0884533612467824.

Sak, M. E., T. Gul, M. S. Evsen, H. E. Soydinc, S. Sak, A. Ozler, and U. Alabalik. 2013. "Fibroblast growth factor-1 expression in the endometrium of patients with repeated implantation failure after in vitro fertilization." *Eur Rev Med Pharmacol Sci* 17 (3):398-402.

Salle, B. L., E. E. Delvin, A. Lapillonne, N. J. Bishop, and F. H. Glorieux. 2000. "Perinatal metabolism of vitamin D." *Am J Clin Nutr* 71 (5 Suppl):1317S-24S.

Sassi, F., C. Tamone, and P. D'Amelio. 2018. "Vitamin D: Nutrient, Hormone, and Immunomodulator." *Nutrients* 10 (11). doi: 10.3390/nu10111656.

Shuler, F. D., J. Hendrix, S. Hodroge, and A. Short. 2013. "Antibiotic-like actions of vitamin D." *W V Med J* 109 (1):22-5.

Siiteri, P. K., J. T. Murai, G. L. Hammond, J. A. Nisker, W. J. Raymoure, and R. W. Kuhn. 1982. "The serum transport of steroid hormones." *Recent Prog Horm Res* 38:457-510.

Song, J., Y. Li, and R. An. 2017. "Vitamin D restores angiogenic balance and decreases tumor necrosis factor-alpha in a rat model of pre-eclampsia." *J Obstet Gynaecol Res* 43 (1):42-49. doi: 10.1111/jog.13186.

Speeckaert, M. M., R. Speeckaert, N. van Geel, and J. R. Delanghe. 2014. "Vitamin D binding protein: a multifunctional protein of clinical importance." *Adv Clin Chem* 63:1-57.

Staples, J., A. L. Ponsonby, and L. Lim. 2010. "Low maternal exposure to ultraviolet radiation in pregnancy, month of birth, and risk of multiple sclerosis in offspring: longitudinal analysis." *Bmj* 340:c1640. doi:10.1136/bmj.c1640.

Steenbock, H. 1924. "The Induction of Growth Promoting and Calcifying Properties in a Ration by Exposure to Light." *Science* 60 (1549):224-5. doi: 10.1126/science.60.1549.224.

Takeyama, K., S. Kitanaka, T. Sato, M. Kobori, J. Yanagisawa, and S. Kato. 1997. "25-Hydroxyvitamin D3 1alpha-hydroxylase and vitamin D synthesis." *Science* 277 (5333):1827-30. doi:10.1126/science.277.5333.1827.

Thacher, T. D., and M. A. Levine. 2017. "CYP2R1 mutations causing vitamin D-deficiency rickets." *J Steroid Biochem Mol Biol* 173:333-336. doi: 10.1016/j.jsbmb.2016.07.014.

Thangamani, S., M. Kim, Y. Son, X. Huang, H. Kim, J. H. Lee, J. Cho, B. Ulrich, H. E. Broxmeyer, and C. H. Kim. 2015. "Cutting edge: progesterone directly upregulates vitamin d receptor gene expression for efficient regulation of T cells by calcitriol." *J Immunol* 194 (3):883-6. doi: 10.4049/jimmunol.1401923.

Tolppanen, A. M., A. Sayers, W. D. Fraser, and D. A. Lawlor. 2012. "Association of serum 25-hydroxyvitamin D(3) and D(2) with academic performance in childhood: findings from a prospective birth cohort." *J Epidemiol Community Health* 66 (12):1137-42. doi: 10.1136/jech-2011-200114.

Tous, M., M. Villalobos, L. Iglesias, S. Fernandez-Barres, and V. Arija. 2019. "Vitamin D status during pregnancy and offspring outcomes: a systematic review and meta-analysis of observational studies." *Eur J Clin Nutr*. doi: 10.1038/s41430-018-0373-x.

Tsai H. C., Norman A. W., 1973. "Studies on the mode of action of calciferol. VI. Effect of 1,25-dihydroxy-vitamin D3 on RNA synthesis in the intestinal mucosa. ." *Biochem Biophys Res Commun* 54 622–627.

Tsuprykov, O., X. Chen, C. F. Hocher, R. Skoblo, Yin Lianghong, and B. Hocher. 2018. "Why should we measure free 25(OH) vitamin D?" *J Steroid Biochem Mol Biol* 180:87-104. doi: 10.1016/j.jsbmb.2017.11.014.

Tuckey, R. C., C. Y. S. Cheng, and A. T. Slominski. 2019. "The serum vitamin D metabolome: What we know and what is still to discover." *J Steroid Biochem Mol Biol* 186:4-21. doi: 10.1016/j.jsbmb.2018.09.003.

Tuohimaa, P., J. H. Wang, S. Khan, M. Kuuslahti, K. Qian, T. Manninen, P. Auvinen, M. Vihinen, and Y. R. Lou. 2013. "Gene expression profiles in human and mouse primary cells provide new insights into the differential actions of vitamin D3 metabolites." *PLoS One* 8 (10):e75338. doi: 10.1371/journal.pone.0075338.

Valdivielso, J. M. 2009. "The physiology of vitamin D receptor activation." *Contrib Nephrol* 163:206-12. doi: 10.1159/000223800.

van Bennekum, A., M. Werder, S. T. Thuahnai, C. H. Han, P. Duong, D. L. Williams, P. Wettstein, G. Schulthess, M. C. Phillips, and H. Hauser. 2005. "Class B scavenger receptor-mediated intestinal absorption of dietary beta-carotene and cholesterol." *Biochemistry* 44 (11):4517-25. doi: 10.1021/bi0484320.

Vanherwegen, A. S., C. Gysemans, and C. Mathieu. 2017. "Vitamin D endocrinology on the cross-road between immunity and metabolism." *Mol Cell Endocrinol* 453:52-67. doi: 10.1016/j.mce.2017.04.018.

Vieth, R. 2004. "Why "Vitamin D" is not a hormone, and not a synonym for 1,25-dihydroxy-vitamin D, its analogs or deltanoids." *J Steroid Biochem Mol Biol* 89-90 (1-5):571-3. doi: 10.1016/j.jsbmb.2004.03.037.

Vigano, P., D. Lattuada, S. Mangioni, L. Ermellino, M. Vignali, E. Caporizzo, P. Panina-Bordignon, M. Besozzi, and A. M. Di Blasio. 2006. "Cycling and early pregnant endometrium as a site of regulated expression of the vitamin D system." *J Mol Endocrinol* 36 (3):415-24. doi: 10.1677/jme.1.01946.

Vrachnis, N., F. M. Malamas, S. Sifakis, P. Tsikouras, and Z. Iliodromiti. 2012. "Immune aspects and myometrial actions of progesterone and CRH in labor." *Clin Dev Immunol* 2012:937618. doi:10.1155/2012/937618.

Wacker, M., and M. F. Holick. 2013. "Sunlight and Vitamin D: A global perspective for health." *Dermatoendocrinol* 5 (1):51-108. doi: 10.4161/derm.24494.

Wang, B., M. Cruz Ithier, N. Parobchak, S. M. Yadava, J. Schulkin, and T. Rosen. 2018. "Vitamin D stimulates multiple microRNAs to inhibit CRH and other pro-labor genes in human placenta." *Endocr Connect*. doi:10.1530/ec-18-0345.

Wang, D., J. Song, and H. Ma. 2018. "An in vitro Experimental Insight into the Osteoblast Responses to Vitamin D3 and Its Metabolites." *Pharmacology* 101 (5-6):225-235. doi: 10.1159/000486446.

Wang, H., W. Chen, D. Li, X. Yin, X. Zhang, N. Olsen, and S. G. Zheng. 2017. "Vitamin D and Chronic Diseases." *Aging Dis* 8 (3):346-353. doi: 10.14336/ad.2016.1021.

Wang, Y., J. Zhu, and H. F. DeLuca. 2012. "Where is the vitamin D receptor?" *Arch Biochem Biophys* 523 (1):123-33. doi:10.1016/j.a bb.20 12.04.001.

Webb, A. R., and M. F. Holick. 1988. "The role of sunlight in the cutaneous production of vitamin D3." *Annu Rev Nutr* 8:375-99. doi:10.1146/ann urev.nu.08.070188.002111.

Wharton, B., and N. Bishop. 2003. "Rickets." *Lancet* 362 (9393):1389-400. doi: 10.1016/S0140-6736(03)14636-3.

Wierzbicka, J. M., M. A. Zmijewski, A. Piotrowska, B. Nedoszytko, M. Lange, R. C. Tuckey, and A. T. Slominski. 2016. "Bioactive forms of vitamin D selectively stimulate the skin analog of the hypothalamus-pituitary-adrenal axis in human epidermal keratinocytes." *Mol Cell Endocrinol* 437:312-322. doi: 10.1016/j.mce.2016.08.006.

Wimalawansa, S. J. 2019. "Vitamin D Deficiency: Effects on Oxidative Stress, Epigenetics, Gene Regulation, and Aging." *Biology (Basel)* 8 (2). doi: 10.3390/biology8020030.

Woeckel, V. J., B. C. van der Eerden, M. Schreuders-Koedam, M. Eijken, and J. P. Van Leeuwen. 2013. "1alpha,25-dihydroxyvitamin D3 stimulates activin A production to fine-tune osteoblast-induced mineralization." *J Cell Physiol* 228 (11):2167-74. doi:10.1002/jc p.243 88.

Wolf, G. 2004. "The discovery of vitamin D: the contribution of Adolf Windaus." *J Nutr* 134 (6):1299-302. doi: 10.1093/jn/134.6.1299.

Young, S. L. 2013. "Oestrogen and progesterone action on endometrium: a translational approach to understanding endometrial receptivity." *Reprod Biomed Online* 27 (5):497-505. doi:10.1016/j.rbmo.2 013.06. 010.

Yu, O. B., and L. A. Arnold. 2016. "Calcitroic Acid-A Review." *ACS Chem Biol* 11 (10):2665-2672. doi: 10.1021/acschembio.6b00569.

Zhu, J. G., J. T. Ochalek, M. Kaufmann, G. Jones, and H. F. Deluca. 2013. "CYP2R1 is a major, but not exclusive, contributor to 25-hydroxyvitamin D production in vivo." *Proc Natl Acad Sci U S A* 110 (39):156 50-5. doi: 10.1073/pnas.1315006110.

In: Vitamin D as Progesterone-Like Hormone ISBN: 978-1-53617-516-5
Editor: Vittorio Unfer

Chapter 2

VITAMIN D AND PROGESTERONE: CRITICAL ANALYSIS OF THE FUNCTIONAL OVERLAP IN IMMUNE REGULATION

Julia Szekeres-Bartho**, *MD, PhD
Department of Medical Biology, Pecs University,
Medical School, Pecs, Hungary

ABSTRACT

Vitamin D is increasingly considered as a key factor in several physiological processes. Studies on vitamin D activities have reported a functional overlap between the former and progesterone, which might be explained by the similar chemical structure of the two molecules. Progesterone is known for its immuno-regulatory properties. For this reason, in this chapter, we have chosen to discuss, in detail, the involvement of vitamin D in immune regulation, focusing on the interaction between the latter and progesterone in this particular physiological context. We cover the role of vitamin D in the innate and adaptive immune response, as well as the impact of this molecule in autoimmune diseases with a special emphasis on type 1 diabetes. Finally,

* Corresponding Author's Email: szekeres.julia@pte.hu.

we give an overview on the molecular mechanisms underlying vitamin D activity, and the relationship between this molecule and the main receptors of immune system, e.g., the Toll-like receptors (TLRs). Finally, its synergic and coordinated activity with progesterone in pro-inflammatory situations will be discussed.

Keywords: vitamin D, immunomodulation, progesterone, PIBF, progesterone-induced Blocking factor

INTRODUCTION

Historically, vitamin D has been recognized for its relevance to calcium homeostasis and bone health. Recently, extra-skeletal effects of vitamin D have also been revealed. These findings were supported by the identification of vitamin D receptor (VDR) and vitamin D-activating enzymes (hydroxylases) in tissues and cells that are not involved in mineral and bone metabolism.

In this regard, vitamin D might affect the function of the immune system through pathways that can be also regulated by several immune cell types. Vitamin D affects antigen-presenting cells (APC) which act as a switch between innate and adaptive immunity, thus is involved in both innate and in adaptive immune responses (Wu et al. 2018). The effect of Vitamin D on monocytes and macrophages have been extensively studied (Baeke et al. 2010, Gombart 2009). Vitamin D in physiological concentrations increases the proliferation of human monocytes (Ohta et al. 1985) and the chemotactic and phagocytic activity of macrophages (Xu et al. 1993). Calcitriol ($1,25(OH)_2D_3$) induces the synthesis of several endogenous antimicrobial peptides, e.g., cathelicidin and defensins in monocytes, neutrophils, and epithelial cells (Gombart, Borregaard, and Koeffler 2005, Liu et al. 2006, Wang et al. 2004). These data underline the importance of vitamin D in the elimination of invading pathogens, by activating a network of innate antimicrobial immune responses.

Vitamin D also acts on the adaptive immune response. VDR and vitamin D-activating enzymes are present in both T and B cells (Provvedini et al., 1983). The activation and subsequent proliferation of T or B cells induces over-expression of VDR. Vitamin D has been shown to modulate the adaptive immune response at several levels (Lemire et al. 1984), among others, by inhibiting T cell proliferation (Rigby, Stacy, and Fanger 1984), and effector functions of both CD4+ and CD8+ T cells (Kongsbak et al. 2013, Mora, Iwata, and von Andrian 2008) as well as by down regulating the production of interleukin-2 (IL-2) and interferon gamma (IFN-γ) (Rigby, Denome, and Fanger 1987, Alroy, Towers, and Freedman 1995, Cippitelli and Santoni 1998). A further evidence that shows the impact of Vitamin D on T cell function is its effect on CD4+ T cell differentiation. The cells of the innate immune system produce various cytokines upon encountering particular pathogens. In presence of these cytokines, naïve T cells (Th0) differentiate into several effector subsets; T helper 1 (Th1), T helper 2 (Th2), T helper 17 (Th17), and regulatory T cells (Treg cells). While Th1 and Th17 cells are involved mainly in immunity against intracellular pathogens and are thought to promote inflammation and autoimmunity, Th2 cells are responsible for humoral immunity and targeting extracellular pathogens. Treg cells assist in the maintenance of self-tolerance and regulate immune responses to prevent excessive and mis-directed actions. Several studies confirm the ability of vitamin D to restrict CD4+ T cell polarization toward the pro-inflammatory Th1 and Th17 cells, and to favour the regulatory Th2 and Treg cell development (Baeke et al. 2010, Mora, Iwata, and von Andrian 2008, Wei and Christakos 2015). Another vitamin D action on the immune system now proven is the influence of this molecule on APC function, primarily on dendritic cells (DC). In particular, some studies have reported the ability of vitamin D to help DCs tolerance program, proposing this molecule as a potential therapeutic option for autoimmune and inflammatory diseases (Wu et al. 2018). The effects of vitamin D on different aspects of immune functions will be discussed in this chapter.

THE IMPACT OF VITAMIN D ON THE INNATE IMMUNE RESPONSE: ANTIMICROBIAL ACTIVITY

The innate immune system acts as a first line of defence against infections, rapidly reacting against invading pathogens. The innate immune system consists of physical barriers (e.g., the skin and the mucosal surfaces), phagocytic and APC (e.g., mast cells, DC or macrophages), inflammatory humoral components (e.g., opsonins), enzymes, e.g., lysozyme, antimicrobial peptides (e.g., cathelicidins) and pattern recognition receptors, e.g., Toll-like receptors (TLRs). Vitamin D is a well-known regulator of innate immunity (Sassi, Tamone, and D'Amelio 2018). $1{,}25(OH)_2D_3$ enhances the production of defensin 2 and the cathelicidin antimicrobial peptide (CAMP) increasing the anti-microbial activity of macrophages and monocytes (Wang et al. 2004, Gombart, Borregaard, and Koeffler 2005, Dai et al. 2010). Moreover $1{,}25(OH)_2D_3$ increases chemotaxis, autophagy, and phago-lysosomal fusion of innate immune cells (Liu et al. 2007, White 2010) confirming the importance of vitamin D in the innate immune response. The exposition of human monocytes to pathogens, such as *M. tuberculosis* and others, up-regulates the expression of CYP27B1 and of VDR, thus enhancing both the cell ability to produce $1{,}25(OH)2D_3$ in the site of infection and to respond to this metabolite.

The effect of vitamin D on macrophages is an important aspect for better understanding its importance in the human innate immune system. An essential function of innate immune system is to directly trigger antimicrobial mechanisms to defend against invading pathogens. Whether macrophages do or do not produce anti-microbial peptides, depends on the cytokine environment. IL-15 has been shown to activate the vitamin D-dependent antimicrobial pathway in macrophages. On the other hand, in the presence of IL-10, these cells respond poorly to vitamin D (Krutzik et al. 2008, Kim et al. 2018). Vitamin D dependent upregulation of CAMP is not limited to monocytes or macrophages, it also concerns keratinocytes, epithelial, intestinal, lung and corneal cells, or trophoblasts (all members of the barrier defence system) (Wei and Christakos 2015). The growing number of studies on the role of vitamin D in infections clearly identifies this molecule as a key player in anti-microbial defence (Mathias et al. 2017, Greulich et al. 2017, Quraishi et al. 2015).

These data show that low vitamin D levels are associated with reduced CAMP production, exposing patients to an increasing risk of infections (Ovunc Hacihamdioglu et al. 2016, Georgieva et al. 2019). Respiratory infections were drastically reduced in vitamin D-treated asthmatic patients, and this effect was related to the increase of CAMP levels, and an altered cytokine pattern (Ramos-Martinez et al. 2018).

Vitamin D deficiency may also potentiate Human Immunodeficiency Virus 1 (HIV-1) disease progression. In this regard, Vitamin D deficiency has been shown to be associated with greater HIV-1 replication in peripheral blood cells. High-dosage of oral vitamin D_3 attenuated HIV-1 replication and increased circulating leukocytes (Sly et al. 2001, Shin et al. 2010).

In conclusion, vitamin D modulates the innate immune system, reinforcing the physical barrier function and increasing the phagocytic capacity of immune cells (Sly et al. 2001, Shin et al. 2010). For all these reasons, the vitamin D treatment of acute or chronic infections can be considered a useful therapeutic option to support anti-microbial defence (Sassi, Tamone, and D'Amelio 2018).

Vitamin D and Toll-Like Receptors

As previously reported, Vitamin D, which plays key roles in the calcium and phosphorus homeostasis also participates in several biological processes including the modulation of immune responses (Jamali et al. 2013, Jamali, Arababadi, and Asadikaram 2013). The correlation between inadequate vitamin D levels and malfunctions of immune responses, has been reported in several immune related diseases (e.g., psoriasis (Kim, Park, and Lee 2010), tuberculosis (Liu et al. 2007), HIV (Campbell and Spector 2012), cardiovascular disease (Adamczak 2017), cancers (Garland et al. 2006), hypersensitivities (Kamen and Tangpricha 2010, Toss and Symreng 1983), type I diabetes (Adamczak et al. 2014), Systemic-Sclerosis (Ionescu and Groseanu 2014) or Crohn's disease (Dionne et al. 2014). Recently several studies have analysed pathways that could be involved in vitamin D-mediated modulation of the immune response.

Pathogens are recognized via the interaction between pattern recognition receptors (PRR) and pathogen associated molecular patterns (PAMPs) or damage associated molecular patterns (DAMPs) (Noroozi et al. 2017). TLRs are a set of immune receptors, belonging to the PRR family (Bagheri et al. 2014). These immune receptors consist of a leucine-rich repeats (LRRs) membrane domain, a hydrophobic transmembrane domain and a toll/interleukin-1 receptor (TIR) domain in the cytoplasm. More than 11 TLRs have been described of which TLR1, 2, 4, 5 and 6 expressed on the cytoplasmic membrane and TLR3, 7, 8 and 9 expressed on the vesicles membrane (Takeda, Kaisho, and Akira 2003). TLR2, to recognize its ligand works in combination with TLR1 and 6 (Bagheri et al. 2014), while, other TLRs recognize their ligands by themselves (Karimi-Googheri and Arababadi 2014, Zare-Bidaki, Tsukiyama-Kohara, and Arababadi 2014). The most important ligands of TLRs are the following: bacterial diacylated lipopeptides/lipoteichoic acid, bacterial lipopeptides, double stranded RNA (dsRNA), lypopolysacharide (LPS), bacterial flagellin, single stranded RNA (ssRNA) and double stranded DNA (dsDNA) (Shahrakyvahed et al. 2014, Takeda, Kaisho, and Akira 2003, Zare-Bidaki, Tsukiyama-Kohara, and Arababadi 2014).

When TLRs recognize PAMP or DAMP, pro-inflammatory transcription factors, e.g., Nuclear Factor kappa-light-chain-enhancer of activated B cells (NF-κB), Activator Protein 1 (AP-1), Interferons (IFNs) etc.. (Karimi-Googheri and Arababadi 2014) are activated.

Considering the important roles played by vitamin D on the expression and functions of immune-related molecules, several groups have investigated the relationship between this molecule and TLRs.

Monocytes exposed to $1,25(OH)_2D_3$ have shown impaired TLR signalling, characterized by the block of NF-kB/p65 translocation to the nucleus and reduced p38 and p42/44 (ERK1/2) stress-activated protein kinase (SAPK) signalling (Sadeghi et al. 2006). These data suggest that vitamin D acts on TLR-signalling. TLRs are fundamental for pathogen-induced immune activation but, in some cases, a dangerous overstimulation can occur. Considering that an excessive activation of immune response might result in hypersensitivity or autoimmunity, the immunomodulatory effects of vitamin D could be an important option to overcome pathologies. To explore this option, the effect of vitamin D on the expression and functions of TLRs in the immune and non-immune cells has been studied. Adams et al. have reported the ability of vitamin D to regulate the expressions of TLR2 and 4 on monocytes of Type 1 Diabetes Mellitus (T1DM) patients (Adams et al. 2009). Several studies have shown that an increased expression of TLRs, especially TLR2 and 4, in pro-inflammatory situations, is associated with autoimmune diseases (Bagheri et al. 2014, Do et al. 2008, Radman et al. 2015). Another interesting work of Do et al., showed the decreased expression of TLR2 and TLR4, as well as an overall improvement of clinical indicators, after vitamin D administration to vitamin D-deficient patients suffering from Behcet's disease (Do et al. 2008). The effect of vitamin D treatment on TLR4, was also evaluated *in vitro* (Helming et al. 2005, Sadeghi et al. 2006, Scherberich et al. 2005).

Vitamin D on the other hand, had no effect on the expression of TLR2 (Li et al. 2002). This fact could be explained with the dual roles, of TLR2 in the induction of immune responses. TLR2 can exert both anti-inflammatory and pro-inflammatory effects (Bagheri et al. 2014), depending on the context, the antigen is presented in.

Several studies have also reported a vitamin D dependent modulation of TLR7 expression (involved in tumour development) (Chen et al. 2016) and of TLR9 (involved in pro-inflammatory bowel diseases) (Dai et al. 2015).

Another TLR regulated by vitamin D, is TLR10, an anti-inflammatory receptor, normally expressed in macrophages. Vitamin D has been shown to induce over-expression of TLR10 in human microglia (Verma and Kim 2016). Several reports show the reduced expression of this receptor in multiple sclerosis, which is in line with the high prevalence of vitamin D deficiency among multiple sclerosis patients.

Alva-Murillo et al., have reported that cholecalciferol, also known as vitamin D3, inhibited the entry of Staphylococcus aureus (S. aureus) to bovine mammary epithelial cells, and reduced the expression of TLR2 in the S. aureus infected cells. Moreover, in this study, vitamin D induced an increased expression of IL-10 in response to S. aureus (Alva-Murillo et al. 2014), showing not only a reduced susceptibility of the cells but also a regulated immune response against these pathogens. The decreased expression of TLR2 and TLR4 in presence of vitamin D, as well as their downstream pro-inflammatory cytokines induced by was also reported by Jeong et al., in human keratinocytes (Jeong et al. 2014).

In presence of pathogens, macrophages are stimulated through TLR and potent bactericidial peptides (β-defensin 4 and cathelicidin) are produced (Gallo et al. 2002). The co-localization of these two peptides within the phagosomes disrupts the cell wall of the ingested bacteria (Liu et al. 2006). β-defensin 4 and cathelicidin synthesis is regulated by the expression of VDR and 1-α-hydroxylase (Liu et al. 2006, Wang et al. 2004), both of them upregulated after the bacterial stimulation of TLR. Indeed, after vitamin D activation, $1{,}25(OH)_2D_3$-VDR-Retinoid X Receptor (RXR) complex formation induces the transcription of genes for β defensin 4 and cathelicidin resulting in the production of these peptides. Obviously, appropriate vitamin D levels are required for this process, since cathelicidin transcription happens only in the presence of vitamin D (Liu et al. 2006).

Vitamin D modulates the expression of TLRs. Kim et al. demonstrated a negative correlation between TLR2 and VDR expression in vitamin D deficient patients and a positive correlation of the above factors in patients with sufficient vitamin D levels (Kim, Park, and Lee 2010). Overall it appears that vitamin D participates in the modulation of several immune factors, responsible of an increased anti-pathogens activity and/or of an immuno-modulation during hypersensitivities states. Several other papers have confirmed the immuno-modulatory effects of vitamin D acting on TLRs in autoimmune diseases (Li et al. 2014, Ojaimi et al. 2013, Golden et al. 2013, Tiosano et al. 2013, Devaraj et al. 2011, Devaraj et al. 2008, Khoo et al. 2011, Choi, Lee, and Sohn 2011) and, even if more investigations are needed to confirm these data, the direction of research seems to be the right one (Arababadi, Nosratabadi, and Asadikaram 2018).

Effect of Vitamin D on Microbiota

A correct balance between microbiota and the host is fundamental for the development, maturation, and proper function of the immune system (D'Amelio and Sassi 2018). Considering the information shown in the previous section, vitamin D could be one of the actors in the relationship between microbiota and the immune system. This assumption is supported by the correlation between low levels of vitamin D with greater gut permeability and, consequently, with a metabolic endotoxemia that induces a low-grade inflammation (Caricilli et al. 2011). Furthermore, the latter can be corrected by vitamin D supplementation (Flanagan et al. 2015). Moreover, dysbiosis in vitamin D depleted- and/or VDR knockout animal models results in metabolic disorders (Su et al. 2016). Studies in transgenic mice, evaluating the effects of VDR overexpression in the intestinal epithelium, revealed an increased resistance to colitis (Liu et al. 2013, Golan et al. 2015) and a decreased mucosal inflammation improving tight junction function (Liu et al. 2013, Du et al. 2015). On the contrary, selective deletion of VDR in the bowel favours a Th1 and Th17 types mucosal infiltration and inflammatory cytokines production resulting in a more severe form of colitis (He et al. 2018).

In line with this, observational studies on humans have shown a strong correlation between low levels of calcidiol ($25(OH)D_3$) and an increased risk of inflammatory bowel disease (IBD) (Loftus and Sandborn 2002, Loftus 2004, Lim, Hanauer, and Li 2005), while high levels of $25(OH)D_3$ decrease the risk of Clostridium difficile infection (Ananthakrishnan et al. 2014). These data are confirmed by a significant reduction of VDR expression in the colonic epithelium of patients affected by IBD in comparison to that of healthy controls (Liu et al. 2013, Du et al. 2015, Schaffler et al. 2018). Nevertheless, human data on the effect of vitamin D supplementation on microbiota in IBD are still controversial (Garg et al. 2018, Luthold et al. 2017, Bashir et al. 2016) and, for this reason, further confirmations are needed.

However, the impact of vitamin D in the microbiome is not limited to the gut. In cystic fibrosis patients, vitamin D insufficiency is associated with an altered composition of the microbiome in the airways (Kanhere et al. 2018). Obviously, considering the complexity of microbiota dynamics, further studies are needed to fully understand the effects of vitamin D in this context.

VITAMIN D AND THE ADAPTIVE IMMUNE SYSTEM

The adaptive immune response consists of mechanisms and molecules that target - self or non self - specific antigens. The adaptive immune response starts when the appropriately presented antigen is recognized by T and B cells. T cell recognize antigens in a Major Histocompatibility Complex (MHC) restricted manner. Cytotoxic T cells expressing foreign antigens, while helper T cells produce cytokines that aid B cell and T cell responses. After recognizing the specific antigen, with the help of T cell derived cytokines, B cells differentiate into antibody producing plasma cells. Memory cells are produced during the immune response.

Vitamin D acts as an immunomodulator molecule, attenuating strong pro-inflammatory reactions (Wei and Christakos 2015, Chun et al. 2014). Vitamin D down-regulates the immune responses mediated by Th1 cells, by inhibiting the production of pro-inflammatory cytokines, (IFN-γ, IL-6, IL-2, and TNF-α) (Carvalho et al. 2017, Xie et al. 2017). Several reports have confirmed the suppressive effect of vitamin D on Th1 responses in various pathological conditions, e.g., spinal tuberculosis (Zhou et al. 2018), uremia (Stubbs et al. 2010), and autoimmune thyroiditis (Drozdenko, Heine, and Worm 2014).

Although it is difficult to compare these studies considering the different doses and clinical conditions, the main conclusion is that $1,25(OH)_2D_3$ acts as an immune modulator, not only by suppressing Th1 cells activation, but also by acting on Th2 cells, Tregs and Th17 cells.

In vitro studies revealed a $1,25(OH)_2D_3$ dependent upregulation of Th2 cells activity (Boonstra et al. 2001, Pichler et al. 2002, Staeva-Vieira and Freedman 2002). Vitamin D-dependent suppression of Th17 and increase of Treg activity has also been reported (Fawaz et al. 2016, Siklar et al. 2016,

Korf et al. 2012) with important consequences in autoimmune diseases (Xie et al. 2017, Mann et al. 2017, Joshi et al. 2011, Chang et al. 2010, Colin et al. 2010, Palmer et al. 2011) as the T1DM (Treiber et al. 2015).

DC direct the differentiation of Th0 cells to an effector cells with pro- or anti-inflammatory properties. According to a recent hypothesis, the immunomodulatory role of vitamin D could be due to the effect of this molecule on the action of $1,25(OH)_2D_3$ on DC (Hu and Wan 2011). In vitro studies on the immune cell differentiation, in presence of $1,25(OH)_2D_3$, have shown the induction of a tolerogenic phenotype. This was characterized by low levels of inflammatory cytokines (e.g., IL-12 and TNF-α) and an increased level of the anti-inflammatory IL-10 as well as by cell differentiation to Treg cells and the apoptosis of autoreactive T cells (Penna and Adorini 2000, Piemonti et al. 2000, Unger et al. 2009, van Halteren et al. 2004). However, the therapeutic use of vitamin D and its metabolites in order to ameliorate the adaptive immune response is not confirmed by a sufficient amount of data.

Vitamin D and Autoimmune Diseases

The growing number of evidences on the immunomodulatory activities of vitamin D as well as the immunological abnormalities in vitamin D deficient individuals and the beneficial effects of its supplementation confirm the immunological importance of this molecule.

Studies in animal models showed that $1,25(OH)_2D_3$ supplementation prevented the initiation and reduced the progression of autoimmune encephalomyelitis (EAE), collagen-induced arthritis (CIA) (Lemire and Archer 1991, Cantorna, Hayes, and DeLuca 1996, 1998), as well as a severe form of enterocolitis (Alhassan Mohammed et al. 2017).

Epidemiological studies revealed a relationship between the increasing latitude (consequently decreasing sunlight exposure) and a higher prevalence of multiple sclerosis (Tao et al. 2016, Laursen et al. 2016, Jelinek et al. 2015), type I diabetes (Mohr et al. 2008, Rosecrans and Dohnal 2014, van der Rhee, de Vries, and Coebergh 2016), and IBD (Szilagyi et al. 2014) suggesting, that there may be a correlation between vitamin D levels and autoimmune diseases. This idea is supported by the fact that subjects born in the northern hemisphere during the period of lower sunshine hours, have an increased risk of developing multiple sclerosis compared to patients born in the months with higher daylight duration (Dobson, Giovannoni, and Ramagopalan 2013). Several studies correlate the dietary intake of vitamin D with the prevalence of autoimmune diseases as rheumatoid arthritis (Song, Bae, and Lee 2012) and T1DM (Zipitis and Akobeng 2008, Dong et al. 2013) yet, there is no evidence for a causative relationship between vitamin D deficiency and these diseases.

Prevention and Treatment of Type 1 Diabetes Mellitus: the Contribution of Vitamin D

T1DM is a chronic autoimmune disease, characterised by degeneration of pancreatic β-cells and impaired insulin production. The chronic autoimmune nature of this pathology induces a steady inflammatory state. T1DM can have a genetic background that favours the disease onset, particularly in childhood (Paschou et al. 2018, Noble and Erlich 2012, Frese et al. 2008). Also, dietary factors seem to be an important trigger for T1DM and, interestingly, vitamin D insufficiency and deficiency have been reported, as strong dietary risk factors for T1DM development.

Some studies have tried to identify the potential molecular background behind the connection between this molecule and T1DM but this is not an easy task, considering the number of proteins (VDR, vitamin D binding protein (VDBP), CYP2R1, CYP27B1, etc.) involved in the vitamin D metabolic pathways and all their potential polymorphisms (Abd-Allah et al. 2014, Ali et al. 2018, Wang et al. 2014, Sahin et al. 2017, Chang et al. 2000, Mukhtar et al. 2017, Panierakis et al. 2009, Tizaoui et al. 2014, Nykjaer et al. 1999, Arnaud and Constans 1993, Bouillon et al. 1981, Ongagna et al. 2001, Ongagna, Pinget, and Belcourt 2005, Thorsen et al. 2014, Blanton et al. 2011, Hussein, Mohamed, and Alghobashy 2012, Cooper et al. 2011,

Ramos-Lopez et al. 2007, Bailey et al. 2007). Data on VDR gene polymorphisms (named Fok-I, Bsm-I, Apa-I and Taq-I) and T1DM susceptibility are inconsistent and inconclusive. Three polymorphisms of the VDBP, able to alter the affinity for 1,25$(OH)_2$D (Arnaud and Constans, 1993) and to affect the serum 1,25$(OH)_2$D levels, have been analysed (Chang et al. 2010), and no correlation between these and T1DM has been confirmed. No association was found between polymorphisms of the two main enzymes involved in the vitamin D metabolic pathway, CYP2R1 and CYP27B1, and T1DM susceptibility in most (Penna and Adorini 2000, Piemonti et al. 2000, Unger et al. 2009, van Halteren et al. 2004), but not all of the studies (Treiber et al. 2015). For all these reasons additional studies are needed to establish or rule out the association between polymorphisms in genes involved in the vitamin D metabolic pathway and T1DM.

The immunomodulatory activity of vitamin D is based on its ability to regulate gene transcription. In context of autoimmune diseases, the ability of vitamin D to downregulate several mechanisms of adaptive immunity is fundamental. As previously reported, vitamin D enhances the maturation macrophages but reduces their ability to present antigens to T cells, by decreasing MHC-II expression (Overbergh et al. 2000, Korf et al. 2012). During DC maturation, the presence of vitamin D favours the formation of the tolerogenic phenotype down regulated MHC expression, and consequently with impaired antigen presenting capacity (Zhang et al. 2015, Adorini, Giarratana, and Penna 2004, Adorini et al. 2003, Ferreira et al. 2015, Piemonti et al. 2000, Gauzzi et al. 2005).

Vitamin D inhibits and/or impairs B cell proliferation and differentiation into plasma cells, the formation of memory B cells and the production of immunoglobulins. (Chen et al. 2007, Lemire et al. 1984, Geldmeyer-Hilt et al. 2011). The ability of vitamin D to address the immune cell differentiation into Th2 and Treg decreases the Th1/Th2 ratio (Piantoni et al. 2015, Tian et al. 2012, Colin et al. 2010, Jeffery et al. 2009) and affects cytokine production. The release of anti-inflammatory cytokines, such as IL-4, IL-10 and TGF-β, is stimulated in the immune cells at the expense of pro-inflammatory cytokine production, including IL-1β, IL-6, IL-12, IL-17, IL-

22, TNF-α and IFN-γ (Korf et al. 2012, Zhang et al. 2015, Gauzzi et al. 2005, Rubtsov et al. 2008, Colin et al. 2010, Jeffery et al. 2009, Heine et al. 2008).

Taken together, vitamin D effects reduce the inflammatory response, and this suggests a therapeutic potential for autoimmune diseases.

Several studies have demonstrated the association between vitamin D and T1DM. A Swedish study reported on significantly lower 25(OH)D_3 concentrations on T1DM patients, compared to the control group (Littorin et al. 2006). Vitamin D deficiency in patients with T1DM has a high prevalence, in comparison to healthy individuals, as shown in several case control studies (Greer et al. 2013, Daga et al. 2012, Federico et al. 2018).

Further confirmations were provided by two studies comparing vitamin D levels of children suffering from T1DM to healthy subjects (Frese et al. 2008, Abd-Allah et al. 2014, Ali et al. 2018). Although a vitamin D deficiency was frequently found in patients with T1DM, the causative role of vitamin deficiency has never been confirmed.

Nevertheless, several hypotheses were proposed. While some studies consider vitamin D deficiency as a risk factor for T1DM, others suggest that a low level of vitamin D is a consequence of physiological and/or behavioural changes due to T1DM. Interestingly, the impact of vitamin D level on T1DM risk appears to be age dependent, and though there is no evidence that low levels of vitamin D during pregnancy increases the risk of T1DM in the offspring, vitamin D supplementation in infancy reduces the risk of T1DM. Indeed, as reported in a meta-analysis, supplementation with vitamin D in childhood after the 7th month is more effective in reducing the risk of T1DM than vitamin D supplementation later in life (Zipitis and Akobeng 2008). This observation has been confirmed by several groups (Dong et al. 2013, Cadario et al. 2015, Zipitis and Akobeng 2008, Stene, Joner, and Norwegian Childhood Diabetes Study 2003).

Studies on young adults (US American active-duty military personnel) have then analysed the impact of vitamin D on T1DM risk. The data show that individuals with low 25(OH)D_3 levels, developed insulin-requiring DM compared to age-matched controls (Gorham et al. 2012, Munger et al. 2013). Taken together, the above findings suggest that vitamin D status can be considered an age dependent environmental risk factor of T1DM.

VITAMIN D AS TREATMENT OF T1DM

In patients with T1DM undergoing insulin therapy, vitamin D supplementation resulted in the preservation of residual pancreatic β-cells function (Felicio et al. 2018, Bogdanou et al. 2017, Li et al. 2009, Ataie-Jafari et al. 2013, Pitocco et al. 2006). In particular, a lower daily insulin dose (DID) was required and higher levels of fasting C-peptide (FCP) were observed in the treated groups (Li et al. 2009, Ataie-Jafari et al. 2013, Pitocco et al. 2006, Mishra et al. 2016). Moreover, Federico et al. have demonstrated a significant reduction of autoimmune phenomena in patients supplemented with $25(OH)D_3$, together with the maintenance of pancreatic β-cells function (Federico et al. 2014). Further interesting results were obtained in the interventional studies conducted by Giri et al., Dehkordi et al. and Bogdanou et al. All of these authors reported on the improvement of glycaemic control in T1DM patients treated with cholecalciferol adjuvant therapy, with a significant decrease in glycated hemoglobin (HbA1C) level after 3 months of vitamin D treatment (Bogdanou et al. 2017, Giri et al. 2017, Dehkordi et al. 2018).

Interventional studies and randomized controlled trials (RCTs) have then shown that insulin therapy, when supplemented with cholecalciferol, has a protective immunological effect both in patients with T1DM of recent onset and in long-term patients. A significant increase in Treg cells in patients supplemented with cholecalciferol (2000 IU) daily for 12 months was observed in the work of Gabbay et al. (Gabbay et al. 2012) and a similar effect was reported with the administration of cholecalciferol (4000 IU) daily for 3 months to T1DM patients (Bogdanou et al. 2017). In a RCT, in addition to the increase of Treg cell numbers, Treiber et al. have demonstrated a cholecalciferol-dependent improvement of Treg suppressor functions (Treiber et al. 2015). Patients supplemented for 12 months with daily 70 IU/kg cholecalciferol has shown a significant increase of Treg cell suppressive capacity in comparison with the placebo group (Treiber et al. 2015). Further studies are needed to establish the optimal vitamin D dosage in T1DM, considering the individual needs of diabetic patients (Rak and Bronkowska 2018).

PROGESTERONE, VITAMIN D AND THE IMMUNE SYSTEM

Steroid hormones (progesterone, estrogen, androgens, mineralocorticoids, vitamin D, etc..) play important roles in the regulation of several biological processes (Kim 2015). The pregnancy-protective progesterone, and vitamin D - that regulates calcium levels - play important roles in immune regulation. Though both bind to their own receptors and exert their specific biological functions, the two steroid hormones cooperate with each other immunoregulation (Miyaura and Iwata 2002). The connection between the two molecules is the VDR, which is induced by progesterone in T cells (Lee, Lydon, and Kim 2012).

Though implantation and labour take place in an inflammatory environment, the rest of gestation requires non-inflammatory conditions, and the failure to suppress inflammation can result in premature pregnancy termination (Hill, Polgar, and Anderson 1995, Raghupathy 1997, Nakashima et al. 2010, Toldi et al. 2011). The main components to immune-adaptation during pregnancy (all influenced by progesterone) include an altered cytokine balance, reduced T cell responses and natural killer cells (NK) activity, as well as increased activity of regulatory T cells (Piccinni et al. 1995, Miyaura and Iwata 2002, Lee et al. 2011, Lee, Lydon, and Kim 2012). Moreover, progesterone is a known regulator of myeloid and DC development, a fundamental step in the immune tolerance process (Butts et al. 2007). As previously reported, vitamin D is also an immune-regulatory steroid hormone that suppresses Th0 maturation into Th1 and Th17 cells, favouring T-cell differentiation to Treg cells.

Considering this overlap in the immune regulation, several studies have evaluated if these two hormones could have a combined synergic action and, recently, Thangamani et al. have reported a more effective T cells regulation by calcitriol when progesterone is present at the same time. Indeed, progesterone increased the T-cell response to calcitriol by three to five times (Thangamani et al. 2015). Another study revealed that the VDR gene expression, induced by progesterone, was directly mediated through the binding of nuclear progesterone receptors to the upstream regulatory region of the VDR gene. This type of sequential regulation allows a more refined immune control since the co-regulation with two hormones allows a tissue-specific and time-dependent immune system management (Kim 2015).

VITAMIN D, PROGESTERONE AND PIBF

Progesterone is crucial for both the establishment and maintenance of pregnancy. During pregnancy, progesterone induces in lymphocytes the expression of a protein called progesterone-induced blocking factor (PIBF), which mediates some of its immunological effects (Szekeres-Bartho et al. 1985). This protein, with a predicted molecular mass of 89 kDa, shows no significant amino acid sequence homology with any known protein (Polgar et al. 2003). The PIBF expression, originally described in pregnancy as the production and secretion of this molecule by lymphocytes, later was reported also for various cell types. While the full-length PIBF is associated with the nucleus (Lachmann et al. 2004, Kim, Lee, and Rhee 2012, Kim and Rhee 2011) and plays a role in cell cycle regulation (Miko et al. 2011, Halasz et al. 2013, Balassa et al. 2018), shorter isoforms produced by alternative splicing are localized in the cytoplasm, and once secreted, act as cytokines.

Among the immunological actions of PIBF there is the effect on the cytokine balance, and on NK activity.

Pregnancy is characterized by a Th2 dominant cytokine production. The Th1/Th2 ratio is lower in the peripheral blood of healthy pregnant women than in that of nonpregnant individuals or in women with pathological pregnancies (Salazar Garcia et al. 2018) and in pregnant mice (Chaouat et al. 1990). The presence of PIBF alters the Th1/Th2 balance favouring a 8–10 times increased production of IL-10, IL-4, and IL-5 (Szekeres-Bartho and Wegmann 1996). The impact of PIBF in a physiological pregnancy is shown by Raghupathy et al. (Raghupathy et al. 2005). They reported that progesterone treated human pregnancy lymphocytes, produce PIBF, followed by a decreased production of Th1 cytokines and an increased production of Th2 cytokines.

PIBF has been shown to alter NK cell cytotoxic activity (Kinsky et al. 1990, Szekeres-Bartho, Kinsky, and Chaouat 1990). In humans, recurrent miscarriages are frequently associated with an increased number of endometrial NK cells (Quenby and Farquharson 2006) and although decidual NK cells do not express nuclear progesterone receptor, they still appear to be affected by PIBF. Faust et al. (Faust et al. 1999) showed that PIBF inhibits the cytotoxicity of peripheral NK cells via a block of degranulation without interfering with target conjugation.

PIBF is present in pregnancy serum as well as in the urine of pregnant women, and its concentration is predictive of the outcome of pregnancy. A study on 500 pregnant women and 80 nonpregnant individuals revealed that during normal human pregnancy the concentration of PIBF in urine samples continuously increases until the 37th gestational week and starts to decrease thereafter, disappearing when labour starts. In women with threatened miscarriage or threatened preterm delivery, urinary PIBF levels remain significantly lower than those of healthy pregnant women of corresponding gestational ages (Polgar et al. 2004). In the urine of patients with preeclampsia, PIBF concentrations are significantly lower than in normal pregnancy, and they correlate with the number of symptoms presented. These data suggest that PIBF production is a characteristic feature of normal pregnancy and that determination of PIBF concentrations in urine or in the serum of pregnant women might be of use for the diagnosis of threatened premature pregnancy termination (Szekeres-Bartho 2018). Taken together, the progesterone-dependent mediator PIBF is an important player in the immunological processes, contributing to the maintenance of pregnancy (Szekeres-Bartho 2018).

In light of this, and considering the functional overlap, in the immunological effects of progesterone and vitamin D, it might be of interest to investigate the potential interactions between vitamin D and PIBF. Earlier we showed that vitamin D inhibits natural cytotoxicity in a concentration-dependent manner (Najbauer, Szekeres-Bartho J Fau - Tigyi, and Tigyi).

Recently, we found that, similarly to progesterone, Vitamin D is able to induce PIBF production in activated human peripheral lymphocytes. PIBF-inducing capacity of vitamin D was significantly higher in lymphocytes obtained during the luteal phase, than in those of the follicular phase, suggesting that vitamin D acts more efficiently in progesterone-primed lymphocytes. In line with this, peripheral lymphocytes treated with increasing concentrations of vitamin D plus a standard concentration of progesterone produced significantly more PIBF than those treated with similar concentrations of vitamin D in the absence of progesterone (manuscript in preparation). These data can be explained and, at the same time, confirmed by the finding that progesterone induces VDR in the lymphocytes (Lee et al. 2011), thus progesterone-treated lymphocytes have more available receptors to bind vitamin D.

References

Abd-Allah, S. H., H. F. Pasha, H. A. Hagrass, and A. A. Alghobashy. 2014. "Vitamin D status and vitamin D receptor gene polymorphisms and susceptibility to type 1 diabetes in Egyptian children." *Gene* 536 (2):430-4. doi: 10.1016/j.gene.2013.12.032.

Adamczak, D. M. 2017. "The Role of Toll-Like Receptors and Vitamin D in Cardiovascular Diseases-A Review." *Int J Mol Sci* 18 (11). doi: 10.3390/ijms18112252.

Adamczak, D. M., J. K. Nowak, M. Frydrychowicz, M. Kaczmarek, and J. Sikora. 2014. "The role of Toll-like receptors and vitamin D in diabetes mellitus type 1–a review." *Scand J Immunol* 80 (2):75-84. doi: 10.1111/sji.12188.

Adams, J. S., S. Ren, P. T. Liu, R. F. Chun, V. Lagishetty, A. F. Gombart, N. Borregaard, R. L. Modlin, and M. Hewison. 2009. "Vitamin d-directed rheostatic regulation of monocyte antibacterial responses." *J Immunol* 182 (7):4289-95. doi: 10.4049/jimmunol.0803736.

Adorini, L., N. Giarratana, and G. Penna. 2004. "Pharmacological induction of tolerogenic dendritic cells and regulatory T cells." *Semin Immunol* 16 (2):127-34. doi: 10.1016/j.smim.2003.12.008.

Adorini, L., G. Penna, N. Giarratana, and M. Uskokovic. 2003. "Tolerogenic dendritic cells induced by vitamin D receptor ligands enhance regulatory T cells inhibiting allograft rejection and autoimmune diseases." *J Cell Biochem* 88 (2):227-33. doi: 10.1002/jcb.10340.

Alhassan Mohammed, H., A. Mirshafiey, H. Vahedi, G. Hemmasi, A. Moussavi Nasl Khameneh, K. Parastouei, and A. A. Saboor-Yaraghi. 2017. "Immunoregulation of Inflammatory and Inhibitory Cytokines by Vitamin D3 in Patients with Inflammatory Bowel Diseases." *Scand J Immunol* 85 (6):386-394. doi: 10.1111/sji.12547.

Ali, R., I. Fawzy, I. Mohsen, and A. Settin. 2018. "Evaluation of vitamin D receptor gene polymorphisms (Fok-I and Bsm-I) in T1DM Saudi children." *J Clin Lab Anal* 32 (5):e22397. doi: 10.1002/jcla.22397.

Alroy, I., T. L. Towers, and L. P. Freedman. 1995. "Transcriptional repression of the interleukin-2 gene by vitamin D3: direct inhibition of NFATp/AP-1 complex formation by a nuclear hormone receptor." *Mol Cell Biol* 15 (10):5789-99.

Alva-Murillo, N., A. D. Tellez-Perez, I. Medina-Estrada, C. Alvarez-Aguilar, A. Ochoa-Zarzosa, and J. E. Lopez-Meza. 2014. "Modulation of the inflammatory response of bovine mammary epithelial cells by cholecalciferol (vitamin D) during Staphylococcus aureus internalization." *Microb Pathog* 77:24-30. doi: 10.1016/j.micpath.201 4.10.006.

Ananthakrishnan, A. N., A. Cagan, V. S. Gainer, S. C. Cheng, T. Cai, P. Szolovits, S. Y. Shaw, S. Churchill, E. W. Karlson, S. N. Murphy, I. Kohane, and K. P. Liao. 2014. "Higher plasma vitamin D is associated with reduced risk of Clostridium difficile infection in patients with inflammatory bowel diseases." *Aliment Pharmacol Ther* 39 (10):1136-42. doi: 10.1111/apt.12706.

Arababadi, M. K., R. Nosratabadi, and G. Asadikaram. 2018. "Vitamin D and toll like receptors." *Life Sci* 203:105-111. doi: 10.1016/j.lfs.2 018. 03.040.

Arnaud, J., and J. Constans. 1993. "Affinity differences for vitamin D metabolites associated with the genetic isoforms of the human serum carrier protein (DBP)." *Hum Genet* 92 (2):183-8.

Ataie-Jafari, A., S. C. Loke, A. B. Rahmat, B. Larijani, F. Abbasi, M. K. Leow, and Z. Yassin. 2013. "A randomized placebo-controlled trial of alphacalcidol on the preservation of beta cell function in children with recent onset type 1 diabetes." *Clin Nutr* 32 (6):911-7. doi: 10.1016/j. clnu.2013.01.012.

Baeke, F., T. Takiishi, H. Korf, C. Gysemans, and C. Mathieu. 2010. "Vitamin D: modulator of the immune system." *Curr Opin Pharmacol* 10 (4):482-96. doi: 10.1016/j.coph.2010.04.001.

Bagheri, V., A. Askari, M. K. Arababadi, and D. Kennedy. 2014. "Can Toll-Like Receptor (TLR) 2 be considered as a new target for immunotherapy against hepatitis B infection?" *Hum Immunol* 75 (6):549-54. doi:10.10 16/j.humimm.2014.02.018.

Bailey, R., J. D. Cooper, L. Zeitels, D. J. Smyth, J. H. Yang, N. M. Walker, E. Hypponen, D. B. Dunger, E. Ramos-Lopez, K. Badenhoop, S. Nejentsev, and J. A. Todd. 2007. "Association of the vitamin D metabolism gene CYP27B1 with type 1 diabetes." *Diabetes* 56 (10):261 6-21. doi: 10.2337/db07-0652.

Balassa, T., G. Berta, L. Jakab, N. Bohonyi, and J. Szekeres-Bartho. 2018. "The effect of the Progesterone-Induced Blocking Factor (PIBF) on E-cadherin expression, cell motility and invasion of primary tumour cell lines." *J Reprod Immunol* 125:8-15. doi: 10.1016/j.jri.2017.10.047.

Bashir, M., B. Prietl, M. Tauschmann, S. I. Mautner, P. K. Kump, G. Treiber, P. Wurm, G. Gorkiewicz, C. Hogenauer, and T. R. Pieber. 2016. "Effects of high doses of vitamin D3 on mucosa-associated gut microbiome vary between regions of the human gastrointestinal tract." *Eur J Nutr* 55 (4):1479-89. doi: 10.1007/s00394-015-0966-2.

Blanton, D., Z. Han, L. Bierschenk, M. V. Linga-Reddy, H. Wang, M. Clare-Salzler, M. Haller, D. Schatz, C. Myhr, J. X. She, C. Wasserfall, and M. Atkinson. 2011. "Reduced serum vitamin D-binding protein levels are associated with type 1 diabetes." *Diabetes* 60 (10):2566-70. doi:10.2 337/db11-0576.

Bogdanou, D., M. Penna-Martinez, N. Filmann, T. L. Chung, Y. Moran-Auth, J. Wehrle, C. Cappel, S. Huenecke, E. Herrmann, U. Koehl, and K. Badenhoop. 2017. "T-lymphocyte and glycemic status after vitamin D treatment in type 1 diabetes: A randomized controlled trial with sequential crossover." *Diabetes Metab Res Rev* 33 (3). doi: 10.1002/ dmrr.2865.

Boonstra, A., F. J. Barrat, C. Crain, V. L. Heath, H. F. Savelkoul, and A. O'Garra. 2001. "1alpha,25-Dihydroxyvitamin d3 has a direct effect on naive CD4(+) T cells to enhance the development of Th2 cells." *J Immunol* 167 (9):4974-80.

Bouillon, R., F. A. Van Assche, H. Van Baelen, W. Heyns, and P. De Moor. 1981. "Influence of the vitamin D-binding protein on the serum concentration of 1,25-dihydroxyvitamin D3. Significance of the free 1,25-dihydroxyvitamin D3 concentration." *J Clin Invest* 67 (3):589-96. doi: 10.1172/JCI110072.

Butts, C. L., S. A. Shukair, K. M. Duncan, E. Bowers, C. Horn, E. Belyavskaya, L. Tonelli, and E. M. Sternberg. 2007. "Progesterone inhibits mature rat dendritic cells in a receptor-mediated fashion." *Int Immunol* 19 (3):287-96. doi: 10.1093/intimm/dxl145.

Cadario, F., S. Savastio, V. Pagliardini, M. Bagnati, M. Vidali, F. Cerutti, I. Rabbone, F. Fontana, R. Lera, V. De Donno, A. Valori, G. Gruden, G. Bona, and G. Bruno. 2015. "Vitamin D levels at birth and risk of type 1 diabetes in childhood: a case-control study." *Acta Diabetol* 52 (6):1077-81. doi: 10.1007/s00592-015-0772-6.

Campbell, G. R., and S. A. Spector. 2012. "Toll-like receptor 8 ligands activate a vitamin D mediated autophagic response that inhibits human immunodeficiency virus type 1." *PLoS Pathog* 8 (11):e1003017. doi: 10.1371/journal.ppat.1003017.

Cantorna, M. T., C. E. Hayes, and H. F. DeLuca. 1996. "1,25-Dihydroxyvitamin D3 reversibly blocks the progression of relapsing encephalomyelitis, a model of multiple sclerosis." *Proc Natl Acad Sci U S A* 93 (15):7861-4.

Cantorna, M. T., C. E. Hayes, and H. F. DeLuca. 1998. "1,25-Dihydroxycholecalciferol inhibits the progression of arthritis in murine models of human arthritis." *J Nutr* 128 (1):68-72. doi: 10.1093/jn/128.1.68.

Caricilli, A. M., P. K. Picardi, L. L. de Abreu, M. Ueno, P. O. Prada, E. R. Ropelle, S. M. Hirabara, A. Castoldi, P. Vieira, N. O. Camara, R. Curi, J. B. Carvalheira, and M. J. Saad. 2011. "Gut microbiota is a key modulator of insulin resistance in TLR 2 knockout mice." *PLoS Biol* 9 (12):e1001212. doi: 10.1371/journal.pbio.1001212.

Carvalho, J. T. G., M. Schneider, L. Cuppari, C. C. Grabulosa, T. Aoike D, Q. Redublo BM, C. Batista M, M. Cendoroglo, R. Maria Moyses, and M. A. Dalboni. 2017. "Cholecalciferol decreases inflammation and improves vitamin D regulatory enzymes in lymphocytes in the uremic environment: A randomized controlled pilot trial." *PLoS One* 12 (6):e0179540. doi: 10.1371/journal.pone.0179540.

Chang, J. H., H. R. Cha, D. S. Lee, K. Y. Seo, and M. N. Kweon. 2010. "1,25-Dihydroxyvitamin D3 inhibits the differentiation and migration of T(H)17 cells to protect against experimental autoimmune encephalomyelitis." *PLoS One* 5 (9):e12925. doi: 10.1371/journal.pone.0012925.

Chang, T. J., H. H. Lei, J. I. Yeh, K. C. Chiu, K. C. Lee, M. C. Chen, T. Y. Tai, and L. M. Chuang. 2000. "Vitamin D receptor gene polymorphisms influence susceptibility to type 1 diabetes mellitus in the Taiwanese population." *Clin Endocrinol (Oxf)* 52 (5):575-80.

Chaouat, G., E. Menu, D. A. Clark, M. Dy, M. Minkowski, and T. G. Wegmann. 1990. "Control of fetal survival in CBA x DBA/2 mice by lymphokine therapy." *J Reprod Fertil* 89 (2):447-58.

Chen, J., L. H. Katz, N. M. Munoz, S. Gu, J. H. Shin, W. S. Jogunoori, M. H. Lee, M. D. Belkin, S. B. Kim, J. C. White, J. Andricovich, A. Tzatsos, S. Li, S. S. Kim, K. Shetty, B. Mishra, A. Rashid, J. S. Lee, and L. Mishra. 2016. "Vitamin D Deficiency Promotes Liver Tumor Growth in Transforming Growth Factor-beta/Smad3-Deficient Mice Through Wnt and Toll-like Receptor 7 Pathway Modulation." *Sci Rep* 6:30217. doi: 10.1038/srep30217.

Chen, S., G. P. Sims, X. X. Chen, Y. Y. Gu, S. Chen, and P. E. Lipsky. 2007. "Modulatory effects of 1,25-dihydroxyvitamin D3 on human B cell differentiation." *J Immunol* 179 (3):1634-47.

Choi, B., E. S. Lee, and S. Sohn. 2011. "Vitamin D3 ameliorates herpes simplex virus-induced Behcet's disease-like inflammation in a mouse model through down-regulation of Toll-like receptors." *Clin Exp Rheumatol* 29 (4 Suppl 67):S13-9.

Chun, R. F., P. T. Liu, R. L. Modlin, J. S. Adams, and M. Hewison. 2014. "Impact of vitamin D on immune function: lessons learned from genome-wide analysis." *Front Physiol* 5:151. doi: 10.3389/fphys. 2014.00151.

Cippitelli, M., and A. Santoni. 1998. "Vitamin D3: a transcriptional modulator of the interferon-gamma gene." *Eur J Immunol* 28 (10):3017-30. doi: 10.1002/(SICI)1521-4141(199810)28:10<3017:AID-IMMU3 017>3.0.CO;2-6.

Colin, E. M., P. S. Asmawidjaja, J. P. van Hamburg, A. M. Mus, M. van Driel, J. M. Hazes, J. P. van Leeuwen, and E. Lubberts. 2010. "1,25-dihydroxyvitamin D3 modulates Th17 polarization and interleukin-22 expression by memory T cells from patients with early rheumatoid arthritis." *Arthritis Rheum* 62 (1):132-42. doi: 10.1002/art.25043.

Cooper, J. D., D. J. Smyth, N. M. Walker, H. Stevens, O. S. Burren, C. Wallace, C. Greissl, E. Ramos-Lopez, E. Hypponen, D. B. Dunger, T. D. Spector, W. H. Ouwehand, T. J. Wang, K. Badenhoop, and J. A. Todd. 2011. "Inherited variation in vitamin D genes is associated with predisposition to autoimmune disease type 1 diabetes." *Diabetes* 60 (5):1624-31. doi: 10.2337/db10-1656.

D'Amelio, P., and F. Sassi. 2018. "Gut Microbiota, Immune System, and Bone." *Calcif Tissue Int* 102 (4):415-425. doi: 10.1007/s00223-017-0331-y.

Daga, R. A., B. A. Laway, Z. A. Shah, S. A. Mir, S. K. Kotwal, and A. H. Zargar. 2012. "High prevalence of vitamin D deficiency among newly diagnosed youth-onset diabetes mellitus in north India." *Arq Bras Endocrinol Metabol* 56 (7):423-8.

Dai, X., K. Sayama, M. Tohyama, Y. Shirakata, Y. Hanakawa, S. Tokumaru, L. Yang, S. Hirakawa, and K. Hashimoto. 2010. "PPARgamma mediates innate immunity by regulating the 1alpha, 25-dihydroxy-vitamin D3 induced hBD-3 and cathelicidin in human keratin-ocytes." *J Dermatol Sci* 60 (3):179-86. doi:10.1016/j.jder msci. 2010.09.008.

Dai, Z. H., B. Tan, H. Yang, O. Wang, J. M. Qian, and H. Lv. 2015. "1,25-hydroxyvitamin D relieves colitis in rats via down-regulation of toll-like receptor 9 expression." *Croat Med J* 56 (6):515-24.

Dehkordi, E. H., V. H. Dehkordi, S. M. R. Fatemi, and M. Zolfaghari. 2018. "Effect of vitamin D supplement therapy on HbA1C and IGF-1 levels in children with type 1 diabetes mellitus and vitamin D deficiency." *Electronic Journal of General Medicine* 15 (4):em69.

Devaraj, S., M. R. Dasu, J. Rockwood, W. Winter, S. C. Griffen, and I. Jialal. 2008. "Increased toll-like receptor (TLR) 2 and TLR4 expression in monocytes from patients with type 1 diabetes: further evidence of a proinflammatory state." *J Clin Endocrinol Metab* 93 (2):578-83. doi: 10.1210/jc.2007-2185.

Devaraj, S., J. M. Yun, C. R. Duncan-Staley, and I. Jialal. 2011. "Low vitamin D levels correlate with the proinflammatory state in type 1 diabetic subjects with and without microvascular complications." *Am J Clin Pathol* 135 (3):429-33. doi: 10.1309/AJCPJGZQX42BIAXL.

Dionne, S., M. R. Calderon, J. H. White, B. Memari, I. Elimrani, B. Adelson, C. Piccirillo, and E. G. Seidman. 2014. "Differential effect of vitamin D on NOD2- and TLR-induced cytokines in Crohn's disease." *Mucosal Immunol* 7 (6):1405-15. doi: 10.1038/mi.2014.30.

Do, J. E., S. Y. Kwon, S. Park, and E. S. Lee. 2008. "Effects of vitamin D on expression of Toll-like receptors of monocytes from patients with Behcet's disease." *Rheumatology (Oxford)* 47 (6):840-8. doi: 10.1093/rheumatology/ken109.

Dobson, R., G. Giovannoni, and S. Ramagopalan. 2013. "The month of birth effect in multiple sclerosis: systematic review, meta-analysis and effect of latitude." *J Neurol Neurosurg Psychiatry* 84 (4):427-32. doi: 10.1136/jnnp-2012-303934.

Dong, J. Y., W. G. Zhang, J. J. Chen, Z. L. Zhang, S. F. Han, and L. Q. Qin. 2013. "Vitamin D intake and risk of type 1 diabetes: a meta-analysis of observational studies." *Nutrients* 5 (9):3551-62. doi:10.3390/nu5093551.

Drozdenko, G., G. Heine, and M. Worm. 2014. "Oral vitamin D increases the frequencies of CD38+ human B cells and ameliorates IL-17-producing T cells." *Exp Dermatol* 23 (2):107-12. doi: 10.1111/exd.12300.

Du, J., Y. Chen, Y. Shi, T. Liu, Y. Cao, Y. Tang, X. Ge, H. Nie, C. Zheng, and Y. C. Li. 2015. "1,25-Dihydroxyvitamin D Protects Intestinal Epithelial Barrier by Regulating the Myosin Light Chain Kinase

Signaling Pathway." *Inflamm Bowel Dis* 21 (11):2495-506. doi: 10.1097/MIB.0000000000000526.

Faust, Z., G. Laskarin, D. Rukavina, and J. Szekeres-Bartho. 1999. "Progesterone-induced blocking factor inhibits degranulation of natural killer cells." *Am J Reprod Immunol* 42 (2):71-5.

Fawaz, L., M. F. Mrad, J. M. Kazan, S. Sayegh, R. Akika, and S. J. Khoury. 2016. "Comparative effect of 25(OH)D3 and 1,25(OH)2D3 on Th17 cell differentiation." *Clin Immunol* 166-167:59-71. doi: 10.1016/j.clim.2016.02.011.

Federico, G., D. Focosi, B. Marchi, E. Randazzo, M. De Donno, F. Vierucci, M. Bugliani, F. Campi, F. Scatena, G. Saggese, C. Mathieu, and P. Marchetti. 2014. "Administering 25-hydroxyvitamin D3 in vitamin D-deficient young type 1A diabetic patients reduces reactivity against islet autoantigens." *Clin Nutr* 33 (6):1153-6. doi:10.1016/j.clnu.2014.01.001.

Federico, G., A. Genoni, A. Puggioni, A. Saba, D. Gallo, E. Randazzo, A. Salvatoni, and A. Toniolo. 2018. "Vitamin D status, enterovirus infection, and type 1 diabetes in Italian children/adolescents." *Pediatr Diabetes* 19 (5):923-929. doi: 10.1111/pedi.12673.

Felicio, K. M., Accb de Souza, J. F. A. Neto, F. T. C. de Melo, C. T. Carvalho, T. P. Arbage, H. A. de Rider Brito, A. S. Peixoto, A. F. de Oliveira, F. de Souza Resende, S. S. Reis, A. R. Motta, H. da Costa Miranda, L. C. Janau, E. S. Yamada, and J. S. Felicio. 2018. "Glycemic Variability and Insulin Needs in Patients with Type 1 Diabetes Mellitus Supplemented with Vitamin D: A Pilot Study Using Continuous Glucose Monitoring System." *Curr Diabetes Rev* 14 (4):395-403. doi: 10.2174/1573399813666170616075013.

Ferreira, G. B., A. S. Vanherwegen, G. Eelen, A. C. Gutierrez, L. Van Lommel, K. Marchal, L. Verlinden, A. Verstuyf, T. Nogueira, M. Georgiadou, F. Schuit, D. L. Eizirik, C. Gysemans, P. Carmeliet, L. Overbergh, and C. Mathieu. 2015. "Vitamin D3 Induces Tolerance in Human Dendritic Cells by Activation of Intracellular Metabolic Pathways." *Cell Rep*. doi: 10.1016/j.celrep.2015.01.013.

Flanagan, P. K., D. Chiewchengchol, H. L. Wright, S. W. Edwards, A. Alswied, J. Satsangi, S. Subramanian, J. M. Rhodes, and B. J. Campbell. 2015. "Killing of Escherichia coli by Crohn's Disease Monocyte-derived Macrophages and Its Enhancement by Hydroxychloroquine and Vitamin D." *Inflamm Bowel Dis* 21 (7):1499-510. doi: 10.1097/MIB.00 00000000000387.

Frese, T., H. Sandholzer, S. Voigt, and R. Voigt. 2008. "Epidemiology of diabetes mellitus in German general practitioners' consultation–results of the SESAM 2-study." *Exp Clin Endocrinol Diabetes* 116 (6):326-8.

Gabbay, M. A., M. N. Sato, C. Finazzo, A. J. Duarte, and S. A. Dib. 2012. "Effect of cholecalciferol as adjunctive therapy with insulin on protective immunologic profile and decline of residual beta-cell function in new-onset type 1 diabetes mellitus." *Arch Pediatr Adolesc Med* 166 (7):601-7. doi: 10.1001/archpediatrics.2012.164.

Gallo, R. L., M. Murakami, T. Ohtake, and M. Zaiou. 2002. "Biology and clinical relevance of naturally occurring antimicrobial peptides." *J Allergy Clin Immunol* 110 (6):823-31.

Garg, M., P. Hendy, J. N. Ding, S. Shaw, G. Hold, and A. Hart. 2018. "The Effect of Vitamin D on Intestinal Inflammation and Faecal Microbiota in Patients with Ulcerative Colitis." *J Crohns Colitis* 12 (8):963-972. doi: 10.1093/ecco-jcc/jjy052.

Garland, C. F., F. C. Garland, E. D. Gorham, M. Lipkin, H. Newmark, S. B. Mohr, and M. F. Holick. 2006. "The role of vitamin D in cancer prevention." *Am J Public Health* 96 (2):252-61. doi: 10.2105/AJPH. 2004.045260.

Gauzzi, M. C., C. Purificato, K. Donato, Y. Jin, L. Wang, K. C. Daniel, A. A. Maghazachi, F. Belardelli, L. Adorini, and S. Gessani. 2005. "Suppressive effect of 1alpha, 25-dihydroxyvitamin D3 on type I IFN-mediated monocyte differentiation into dendritic cells: impairment of functional activities and chemotaxis." *J Immunol* 174 (1):270-6.

Geldmeyer-Hilt, K., G. Heine, B. Hartmann, R. Baumgrass, A. Radbruch, and M. Worm. 2011. "1,25-dihydroxyvitamin D3 impairs NF-kappaB activation in human naive B cells." *Biochem Biophys Res Commun* 407 (4):699-702. doi: 10.1016/j.bbrc.2011.03.078.

Georgieva, V., W. Kamolvit, M. Herthelius, P. Luthje, A. Brauner, and M. Chromek. 2019. "Association between vitamin D, antimicrobial peptides and urinary tract infection in infants and young children." *Acta Paediatr* 108 (3):551-556. doi: 10.1111/apa.14499.

Giri, D., D. Pintus, G. Burnside, A. Ghatak, F. Mehta, P. Paul, and S. Senniappan. 2017. "Treating vitamin D deficiency in children with type I diabetes could improve their glycaemic control." *BMC Res Notes* 10 (1):465. doi: 10.1186/s13104-017-2794-3.

Golan, M. A., W. Liu, Y. Shi, L. Chen, J. Wang, T. Liu, and Y. C. Li. 2015. "Transgenic Expression of Vitamin D Receptor in Gut Epithelial Cells Ameliorates Spontaneous Colitis Caused by Interleukin-10 Deficiency." *Dig Dis Sci* 60 (7):1941-7. doi: 10.1007/s10620-015-3634-8.

Golden, G. A., T. A. Wyatt, D. J. Romberger, D. Reiff, M. McCaskill, C. Bauer, A. M. Gleason, and J. A. Poole. 2013. "Vitamin D treatment modulates organic dust-induced cellular and airway inflammatory consequences." *J Biochem Mol Toxicol* 27 (1):77-86. doi: 10.1002/jbt. 21467.

Gombart, A. F. 2009. "The vitamin D-antimicrobial peptide pathway and its role in protection against infection." *Future Microbiol* 4 (9):1151-65. doi: 10.2217/fmb.09.87.

Gombart, A. F., N. Borregaard, and H. P. Koeffler. 2005. "Human cathelicidin antimicrobial peptide (CAMP) gene is a direct target of the vitamin D receptor and is strongly up-regulated in myeloid cells by 1,25-dihydroxyvitamin D3." *FASEB J* 19 (9):1067-77. doi: 10.1096/fj.04-3284com.

Gorham, E. D., C. F. Garland, A. A. Burgi, S. B. Mohr, K. Zeng, H. Hofflich, J. J. Kim, and C. Ricordi. 2012. "Lower prediagnostic serum 25-hydroxyvitamin D concentration is associated with higher risk of insulin-requiring diabetes: a nested case-control study." *Diabetologia* 55 (12):3224-7. doi: 10.1007/s00125-012-2709-8.

Greer, R. M., S. L. Portelli, B. S. Hung, G. J. Cleghorn, S. K. McMahon, J. A. Batch, and L. S. Conwell. 2013. "Serum vitamin D levels are lower in Australian children and adolescents with type 1 diabetes than in children without diabetes." *Pediatr Diabetes* 14 (1):31-41. doi: 10.1111/j.1399-5448.2012.00890.x.

Greulich, T., W. Regner, M. Branscheidt, C. Herr, A. R. Koczulla, C. F. Vogelmeier, and R. Bals. 2017. "Altered blood levels of vitamin D, cathelicidin and parathyroid hormone in patients with sepsis-a pilot study." *Anaesth Intensive Care* 45 (1):36-45.

Halasz, M., B. Polgar, G. Berta, L. Czimbalek, and J. Szekeres-Bartho. 2013. "Progesterone-induced blocking factor differentially regulates trophoblast and tumor invasion by altering matrix metalloproteinase activity." *Cell Mol Life Sci* 70 (23):4617-30. doi: 10.1007/s00018-013-1404-3.

He, L., T. Liu, Y. Shi, F. Tian, H. Hu, D. K. Deb, Y. Chen, M. Bissonnette, and Y. C. Li. 2018. "Gut Epithelial Vitamin D Receptor Regulates Microbiota-Dependent Mucosal Inflammation by Suppressing Intestinal Epithelial Cell Apoptosis." *Endocrinology* 159 (2):967-979. doi:10.1210/en. 2017-00748.

Heine, G., U. Niesner, H. D. Chang, A. Steinmeyer, U. Zugel, T. Zuberbier, A. Radbruch, and M. Worm. 2008. "1,25-dihydroxyvitamin D(3) promotes IL-10 production in human B cells." *Eur J Immunol* 38 (8):2210-8. doi: 10.1002/eji.200838216.

Helming, L., J. Bose, J. Ehrchen, S. Schiebe, T. Frahm, R. Geffers, M. Probst-Kepper, R. Balling, and A. Lengeling. 2005. "1alpha, 25-Dihydroxyvitamin D3 is a potent suppressor of interferon gamma-mediated macrophage activation." *Blood* 106 (13):4351-8. doi:10.1182/blood-2005-03-1029.

Hill, J. A., K. Polgar, and D. J. Anderson. 1995. "T-helper 1-type immunity to trophoblast in women with recurrent spontaneous abortion." *JAMA* 273 (24):1933-6.

Hu, J., and Y. Wan. 2011. "Tolerogenic dendritic cells and their potential applications." *Immunology* 132 (3):307-14. doi: 10.1111/j.1365-2567. 2010.03396.x.

Hussein, A. G., R. H. Mohamed, and A. A. Alghobashy. 2012. "Synergism of CYP2R1 and CYP27B1 polymorphisms and susceptibility to type 1 diabetes in Egyptian children." *Cell Immunol* 279 (1):42-5. doi: 10.1016/j.cellimm.2012.08.006.

Ionescu, R., and L. Groseanu. 2014. "Innate Immunity in Systemic Sclerosis – Role Of Toll-Like Receptors, Interferon, And The Potential Impact of Vitamin D." *European Medical Journal Rheumatology* 1:40-47.

Jamali, Z., M. K. Arababadi, and G. Asadikaram. 2013. "Serum levels of IL-6, IL-10, IL-12, IL-17 and IFN-gamma and their association with markers of bone metabolism in vitamin D-deficient female students." *Inflammation* 36 (1):164-8. doi: 10.1007/s10753-012-9531-9.

Jamali, Z., G. Asadikaram, M. Mahmoodi, A. Sayadi, A. Jamalizadeh, M. Saleh-Moghadam, M. Sirati-Sabt, and M. K. Arababadi. 2013. "Vitamin D status in female students and its relation to calcium metabolism markers, lifestyles, and polymorphism in vitamin D receptor." *Clin Lab* 59 (3-4):407-13.

Jeffery, L. E., F. Burke, M. Mura, Y. Zheng, O. S. Qureshi, M. Hewison, L. S. Walker, D. A. Lammas, K. Raza, and D. M. Sansom. 2009. "1,25-Dihydroxyvitamin D3 and IL-2 combine to inhibit T cell production of inflammatory cytokines and promote development of regulatory T cells expressing CTLA-4 and FoxP3." *J Immunol* 183 (9):5458-67. doi: 10.4049/jimmunol.0803217.

Jelinek, G. A., C. H. Marck, T. J. Weiland, N. Pereira, D. M. van der Meer, and E. J. Hadgkiss. 2015. "Latitude, sun exposure and vitamin D supplementation: associations with quality of life and disease outcomes in a large international cohort of people with multiple sclerosis." *BMC Neurol* 15:132. doi: 10.1186/s12883-015-0394-1.

Jeong, M. S., J. Y. Kim, H. I. Lee, and S. J. Seo. 2014. "Calcitriol May Down-Regulate mRNA Over-Expression of Toll-Like Receptor-2 and -4, LL-37 and Proinflammatory Cytokines in Cultured Human Keratinocytes." *Ann Dermatol* 26 (3):296-302. doi: 10.5021/ad.2014.26.3.296.

Joshi, S., L. C. Pantalena, X. K. Liu, S. L. Gaffen, H. Liu, C. Rohowsky-Kochan, K. Ichiyama, A. Yoshimura, L. Steinman, S. Christakos, and S. Youssef. 2011. "1,25-dihydroxyvitamin D(3) ameliorates Th17 autoimmunity via transcriptional modulation of interleukin-17A." *Mol Cell Biol* 31 (17):3653-69. doi: 10.1128/MCB.05020-11.

Kamen, D. L., and V. Tangpricha. 2010. "Vitamin D and molecular actions on the immune system: modulation of innate and autoimmunity." *J Mol Med (Berl)* 88 (5):441-50. doi: 10.1007/s00109-010-0590-9.

Kanhere, M., J. He, B. Chassaing, T. R. Ziegler, J. A. Alvarez, E. A. Ivie, L. Hao, J. Hanfelt, A. T. Gewirtz, and V. Tangpricha. 2018. "Bolus Weekly Vitamin D3 Supplementation Impacts Gut and Airway Microbiota in Adults With Cystic Fibrosis: A Double-Blind, Randomized, Placebo-Controlled Clinical Trial." *J Clin Endocrinol Metab* 103 (2):564-574. doi: 10.1210/jc.2017-01983.

Karimi-Googheri, M., and M. K. Arababadi. 2014. "TLR3 plays significant roles against hepatitis B virus." *Mol Biol Rep* 41 (5):3279-86. doi: 10.1007/s11033-014-3190-x.

Khoo, A. L., L. Y. Chai, H. J. Koenen, B. J. Kullberg, I. Joosten, A. J. van der Ven, and M. G. Netea. 2011. "1,25-dihydroxyvitamin D3 modulates cytokine production induced by Candida albicans: impact of seasonal variation of immune responses." *J Infect Dis* 203 (1):122-30. doi:10.1 093/infdis/jiq008.

Kim, C. H. 2015. "A functional relay from progesterone to vitamin D in the immune system." *DNA Cell Biol* 34 (6):379-82. doi: 10.1089/dna.2015. 2857.

Kim, E. W., R. M. B. Teles, S. Haile, P. T. Liu, and R. L. Modlin. 2018. "Vitamin D status contributes to the antimicrobial activity of macrophages against Mycobacterium leprae." *PLoS Negl Trop Dis* 12 (7):e0006608. doi: 10.1371/journal.pntd.0006608.

Kim, K., K. Lee, and K. Rhee. 2012. "CEP90 is required for the assembly and centrosomal accumulation of centriolar satellites, which is essential for primary cilia formation." *PLoS One* 7 (10):e48196. doi:10.1371/ journal.pone.0048196.

Kim, K., and K. Rhee. 2011. "The pericentriolar satellite protein CEP90 is crucial for integrity of the mitotic spindle pole." *J Cell Sci* 124 (Pt 3):338-47. doi: 10.1242/jcs.078329.

Kim, S. K., S. Park, and E. S. Lee. 2010. "Toll-like receptors and antimicrobial peptides expressions of psoriasis: correlation with serum vitamin D level." *J Korean Med Sci* 25 (10):1506-12. doi:10.3346/jkms.2010.25.10.1506.

Kinsky, R., G. Delage, N. Rosin, M. N. Thang, M. Hoffmann, and G. Chaouat. 1990. "A murine model of NK cell mediated resorption." *Am J Reprod Immunol* 23 (3):73-7.

Kongsbak, M., T. B. Levring, C. Geisler, and M. R. von Essen. 2013. "The vitamin d receptor and T cell function." *Front Immunol* 4:148. doi: 10.3389/fimmu.2013.00148.

Korf, H., M. Wenes, B. Stijlemans, T. Takiishi, S. Robert, M. Miani, D. L. Eizirik, C. Gysemans, and C. Mathieu. 2012. "1,25-Dihydroxyvitamin D3 curtails the inflammatory and T cell stimulatory capacity of macrophages through an IL-10-dependent mechanism." *Immunobiology* 217 (12):1292-300. doi: 10.1016/j.imbio.2012.07.018.

Krutzik, S. R., M. Hewison, P. T. Liu, J. A. Robles, S. Stenger, J. S. Adams, and R. L. Modlin. 2008. "IL-15 links TLR2/1-induced macrophage differentiation to the vitamin D-dependent antimicrobial pathway." *J Immunol* 181 (10):7115-20.

Lachmann, M., D. Gelbmann, E. Kalman, B. Polgar, M. Buschle, A. Von Gabain, J. Szekeres-Bartho, and E. Nagy. 2004. "PIBF (progesterone induced blocking factor) is overexpressed in highly proliferating cells and associated with the centrosome." *Int J Cancer* 112 (1):51-60. doi: 10.1002/ijc.20326.

Laursen, J. H., H. B. Sondergaard, P. S. Sorensen, F. Sellebjerg, and A. B. Oturai. 2016. "Vitamin D supplementation reduces relapse rate in relapsing-remitting multiple sclerosis patients treated with natalizumab." *Mult Scler Relat Disord* 10:169-173. doi:10.1016/j.msard.2016.10.005.

Lee, J. H., J. P. Lydon, and C. H. Kim. 2012. "Progesterone suppresses the mTOR pathway and promotes generation of induced regulatory T cells with increased stability." *Eur J Immunol* 42 (10):2683-96. doi: 10.1002/eji.201142317.

Lee, J. H., B. Ulrich, J. Cho, J. Park, and C. H. Kim. 2011. "Progesterone promotes differentiation of human cord blood fetal T cells into T regulatory cells but suppresses their differentiation into Th17 cells." *J Immunol* 187 (4):1778-87. doi: 10.4049/jimmunol.1003919.

Lemire, J. M., J. S. Adams, R. Sakai, and S. C. Jordan. 1984. "1 alpha,25-dihydroxyvitamin D3 suppresses proliferation and immunoglobulin production by normal human peripheral blood mononuclear cells." *J Clin Invest* 74 (2):657-61. doi: 10.1172/JCI111465.

Lemire, J. M., and D. C. Archer. 1991. "1,25-dihydroxyvitamin D3 prevents the in vivo induction of murine experimental autoimmune encephalomyelitis." *J Clin Invest* 87 (3):1103-7. doi: 10.1172/JCI11 5072.

Li, C., Y. Wang, L. Gao, J. Zhang, J. Shao, S. Wang, W. Feng, X. Wang, M. Li, and Z. Chang. 2002. "Expression of toll-like receptors 2 and 4 and CD14 during differentiation of HL-60 cells induced by phorbol 12-myristate 13-acetate and 1 alpha, 25-dihydroxy-vitamin D(3)." *Cell Growth Differ* 13 (1):27-38.

Li, P., X. Xu, E. Cao, B. Yu, W. Li, M. Fan, M. Huang, L. Shi, R. Zeng, X. Su, and Y. Shi. 2014. "Vitamin D deficiency causes defective resistance to Aspergillus fumigatus in mice via aggravated and sustained inflammation." *PLoS One* 9 (6):e99805. doi:10.1371/journal.pone.0 099805.

Li, X., L. Liao, X. Yan, G. Huang, J. Lin, M. Lei, X. Wang, and Z. Zhou. 2009. "Protective effects of 1-alpha-hydroxyvitamin D3 on residual beta-cell function in patients with adult-onset latent autoimmune diabetes (LADA)." *Diabetes Metab Res Rev* 25 (5):411-6. doi: 10.10 02/dmrr.977.

Lim, W. C., S. B. Hanauer, and Y. C. Li. 2005. "Mechanisms of disease: vitamin D and inflammatory bowel disease." *Nat Clin Pract Gastroenterol Hepatol* 2 (7):308-15. doi: 10.1038/ncpgasthep0215.

Littorin, B., P. Blom, A. Scholin, H. J. Arnqvist, G. Blohme, J. Bolinder, A. Ekbom-Schnell, J. W. Eriksson, S. Gudbjornsdottir, L. Nystrom, J. Ostman, and G. Sundkvist. 2006. "Lower levels of plasma 25-hydroxyvitamin D among young adults at diagnosis of autoimmune type 1 diabetes compared with control subjects: results from the nationwide Diabetes Incidence Study in Sweden (DISS)." *Diabetologia* 49 (12):284 7-52. doi: 10.1007/s00125-006-0426-x.

Liu, P. T., S. Stenger, H. Li, L. Wenzel, B. H. Tan, S. R. Krutzik, M. T. Ochoa, J. Schauber, K. Wu, C. Meinken, D. L. Kamen, M. Wagner, R. Bals, A. Steinmeyer, U. Zugel, R. L. Gallo, D. Eisenberg, M. Hewison, B. W. Hollis, J. S. Adams, B. R. Bloom, and R. L. Modlin. 2006. "Toll-like receptor triggering of a vitamin D-mediated human antimicrobial response." *Science* 311 (5768):1770-3. doi: 10.1126/science.1123933.

Liu, P. T., S. Stenger, D. H. Tang, and R. L. Modlin. 2007. "Cutting edge: vitamin D-mediated human antimicrobial activity against Mycobacterium tuberculosis is dependent on the induction of cathelicidin." *J Immunol* 179 (4):2060-3.

Liu, W., Y. Chen, M. A. Golan, M. L. Annunziata, J. Du, U. Dougherty, J. Kong, M. Musch, Y. Huang, J. Pekow, C. Zheng, M. Bissonnette, S. B. Hanauer, and Y. C. Li. 2013. "Intestinal epithelial vitamin D receptor signaling inhibits experimental colitis." *J Clin Invest* 123 (9):3983-96. doi: 10.1172/JCI65842.

Loftus, E. V., Jr. 2004. "Clinical epidemiology of inflammatory bowel disease: Incidence, prevalence, and environmental influences." *Gastroenterology* 126 (6):1504-17.

Loftus, E. V., Jr., and W. J. Sandborn. 2002. "Epidemiology of inflammatory bowel disease." *Gastroenterol Clin North Am* 31 (1):1-20.

Luthold, R. V., G. R. Fernandes, A. C. Franco-de-Moraes, L. G. Folchetti, and S. R. Ferreira. 2017. "Gut microbiota interactions with the immunomodulatory role of vitamin D in normal individuals." *Metabolism* 69:76-86. doi: 10.1016/j.metabol.2017.01.007.

Mann, E. H., T. R. Ho, P. E. Pfeffer, N. C. Matthews, E. Chevretton, I. Mudway, F. J. Kelly, and C. M. Hawrylowicz. 2017. "Vitamin D Counteracts an IL-23-Dependent IL-17A(+)IFN-gamma (+) Response Driven by Urban Particulate Matter." *Am J Respir Cell Mol Biol* 57 (3):355-366. doi: 10.1165/rcmb.2016-0409OC.

Mathias, E., V. Tangpricha, A. Sarnaik, A. Farooqi, and U. Sethuraman. 2017. "Association of vitamin D with cathelicidin and vitamin D binding protein in pediatric sepsis." *J Clin Transl Endocrinol* 10:36-38. doi: 10.1016/j.jcte.2017.11.001.

Miko, E., M. Halasz, B. Jericevic-Mulac, L. Wicherek, P. Arck, G. Arato, J. Skret Magierlo, D. Rukavina, and J. Szekeres-Bartho. 2011. "Progesterone-induced blocking factor (PIBF) and trophoblast invasiveness." *J Reprod Immunol* 90 (1):50-7. doi: 10.1016/j.jri. 2011.03.005.

Mishra, A., D. Dayal, N. Sachdeva, and S. V. Attri. 2016. "Effect of 6-months' vitamin D supplementation on residual beta cell function in children with type 1 diabetes: a case control interventional study." *J Pediatr Endocrinol Metab* 29 (4):395-400. doi: 10.1515/jpem-2015-0088.

Miyaura, H., and M. Iwata. 2002. "Direct and indirect inhibition of Th1 development by progesterone and glucocorticoids." *J Immunol* 168 (3):1087-94.

Mohr, S. B., C. F. Garland, E. D. Gorham, and F. C. Garland. 2008. "The association between ultraviolet B irradiance, vitamin D status and incidence rates of type 1 diabetes in 51 regions worldwide." *Diabetologia* 51 (8):1391-8. doi: 10.1007/s00125-008-1061-5.

Mora, J. R., M. Iwata, and U. H. von Andrian. 2008. "Vitamin effects on the immune system: vitamins A and D take centre stage." *Nat Rev Immunol* 8 (9):685-98. doi: 10.1038/nri2378.

Mukhtar, M., A. Batool, A. Wajid, and I. Qayyum. 2017. "Vitamin D Receptor Gene Polymorphisms Influence T1D Susceptibility among Pakistanis." *Int J Genomics* 2017:4171254. doi: 10.1155/2017/4171254.

Munger, K. L., L. I. Levin, J. Massa, R. Horst, T. Orban, and A. Ascherio. 2013. "Preclinical serum 25-hydroxyvitamin D levels and risk of type 1 diabetes in a cohort of US military personnel." *Am J Epidemiol* 177 (5):411-9. doi: 10.1093/aje/kws243.

Najbauer, J., G. J. Szekeres-Bartho J Fau - Tigyi, and G. J. Tigyi. "Modulation of cell-cell and cell-antigen interactions by 1,25-dihydroxyvitamin D3 and vitamin D3 sulfate in vitro: a study on pregnancy lymphocytes and hybridoma cells." (0165-2478 (Print)).

Nakashima, A., M. Ito, T. Shima, N. D. Bac, T. Hidaka, and S. Saito. 2010. "Accumulation of IL-17-positive cells in decidua of inevitable abortion cases." *Am J Reprod Immunol* 64 (1):4-11. doi: 10.1111/j.1600-0897.2010.00812.x.

Noble, J. A., and H. A. Erlich. 2012. "Genetics of type 1 diabetes." *Cold Spring Harb Perspect Med* 2 (1):a007732. doi: 10.1101/cshperspect.a007732.

Noroozi, S., H. A. E. Meimand, M. K. Arababadi, N. Nakhaee, and G. Asadikaram. 2017. "The Effects of IFN-beta 1a on the Expression of Inflammasomes and Apoptosis-Associated Speck-Like Proteins in Multiple Sclerosis Patients." *Mol Neurobiol* 54 (4):3031-3037. doi: 10.1007/s12035-016-9864-8.

Nykjaer, A., D. Dragun, D. Walther, H. Vorum, C. Jacobsen, J. Herz, F. Melsen, E. I. Christensen, and T. E. Willnow. 1999. "An endocytic pathway essential for renal uptake and activation of the steroid 25-(OH) vitamin D3." *Cell* 96 (4):507-15.

Ohta, M., T. Okabe, K. Ozawa, A. Urabe, and F. Takaku. 1985. "1 alpha,25-Dihydroxyvitamin D3 (calcitriol) stimulates proliferation of human circulating monocytes in vitro." *FEBS Lett* 185 (1):9-13.

Ojaimi, S., N. A. Skinner, B. J. Strauss, V. Sundararajan, I. Woolley, and K. Visvanathan. 2013. "Vitamin D deficiency impacts on expression of toll-like receptor-2 and cytokine profile: a pilot study." *J Transl Med* 11:176. doi: 10.1186/1479-5876-11-176.

Ongagna, J. C., M. C. Kaltenbacher, R. Sapin, M. Pinget, and A. Belcourt. 2001. "The HLA-DQB alleles and amino acid variants of the vitamin D-binding protein in diabetic patients in Alsace." *Clin Biochem* 34 (1):59-63.

Ongagna, J. C., M. Pinget, and A. Belcourt. 2005. "Vitamin D-binding protein gene polymorphism association with IA-2 autoantibodies in type 1 diabetes." *Clin Biochem* 38 (5):415-9. doi: 10.1016/j.clinbiochem.2004.12.013.

Overbergh, L., B. Decallonne, D. Valckx, A. Verstuyf, J. Depovere, J. Laureys, O. Rutgeerts, R. Saint-Arnaud, R. Bouillon, and C. Mathieu. 2000. "Identification and immune regulation of 25-hydroxyvitamin D-1-alpha-hydroxylase in murine macrophages." *Clin Exp Immunol* 120 (1):139-46.

Ovunc Hacihamdioglu, D., D. Altun, B. Hacihamdioglu, F. Cekmez, G. Aydemir, M. Kul, T. Muftuoglu, S. Suleymanoglu, and F. Karademir. 2016. "The Association between Serum 25-Hydroxy Vitamin D Level and Urine Cathelicidin in Children with a Urinary Tract Infection." *J Clin Res Pediatr Endocrinol* 8 (3):325-9. doi: 10.4274/jcrpe.2563.

Palmer, M. T., Y. K. Lee, C. L. Maynard, J. R. Oliver, D. D. Bikle, A. M. Jetten, and C. T. Weaver. 2011. "Lineage-specific effects of 1,25-dihydroxyvitamin D(3) on the development of effector CD4 T cells." *J Biol Chem* 286 (2):997-1004. doi: 10.1074/jbc.M110.163790.

Panierakis, C., G. Goulielmos, D. Mamoulakis, E. Petraki, E. Papavasiliou, and E. Galanakis. 2009. "Vitamin D receptor gene polymorphisms and susceptibility to type 1 diabetes in Crete, Greece." *Clin Immunol* 133 (2):276-81. doi: 10.1016/j.clim.2009.08.004.

Paschou, S. A., N. Papadopoulou-Marketou, G. P. Chrousos, and C. Kanaka-Gantenbein. 2018. "On type 1 diabetes mellitus pathogenesis." *Endocr Connect* 7 (1):R38-R46. doi: 10.1530/EC-17-0347.

Penna, G., and L. Adorini. 2000. "1 Alpha,25-dihydroxyvitamin D3 inhibits differentiation, maturation, activation, and survival of dendritic cells leading to impaired alloreactive T cell activation." *J Immunol* 164 (5):2405-11.

Piantoni, S., L. Andreoli, M. Scarsi, A. Zanola, F. Dall'Ara, C. Pizzorni, M. Cutolo, P. Airo, and A. Tincani. 2015. "Phenotype modifications of T-cells and their shift toward a Th2 response in patients with systemic lupus erythematosus supplemented with different monthly regimens of vitamin D." *Lupus* 24 (4-5):490-8. doi: 10.1177/0961203314559090.

Piccinni, M. P., M. G. Giudizi, R. Biagiotti, L. Beloni, L. Giannarini, S. Sampognaro, P. Parronchi, R. Manetti, F. Annunziato, C. Livi, and et al. 1995. "Progesterone favors the development of human T helper cells producing Th2-type cytokines and promotes both IL-4 production and membrane CD30 expression in established Th1 cell clones." *J Immunol* 155 (1):128-33.

Pichler, J., M. Gerstmayr, Z. Szepfalusi, R. Urbanek, M. Peterlik, and M. Willheim. 2002. "1 alpha,25(OH)2D3 inhibits not only Th1 but also Th2 differentiation in human cord blood T cells." *Pediatr Res* 52 (1):12-8. doi: 10.1203/00006450-200207000-00005.

Piemonti, L., P. Monti, M. Sironi, P. Fraticelli, B. E. Leone, E. Dal Cin, P. Allavena, and V. Di Carlo. 2000. "Vitamin D3 affects differentiation, maturation, and function of human monocyte-derived dendritic cells." *J Immunol* 164 (9):4443-51.

Pitocco, D., A. Crino, E. Di Stasio, S. Manfrini, C. Guglielmi, S. Spera, G. B. Anguissola, N. Visalli, C. Suraci, M. C. Matteoli, I. P. Patera, M. G. Cavallo, C. Bizzarri, P. Pozzilli, and Imdiab Group. 2006. "The effects of calcitriol and nicotinamide on residual pancreatic beta-cell function in patients with recent-onset Type 1 diabetes (IMDIAB XI)." *Diabet Med* 23 (8):920-3. doi: 10.1111/j.1464-5491.2006.01921.x.

Polgar, B., G. Kispal, M. Lachmann, C. Paar, E. Nagy, P. Csere, E. Miko, L. Szereday, P. Varga, and J. Szekeres-Bartho. 2003. "Molecular cloning and immunologic characterization of a novel cDNA coding for progesterone-induced blocking factor." *J Immunol* 171 (11):5956-63.

Polgar, B., E. Nagy, E. Miko, P. Varga, and J. Szekeres-Bartho. 2004. "Urinary progesterone-induced blocking factor concentration is related to pregnancy outcome." *Biol Reprod* 71 (5):1699-705. doi: 10.1095/biolreprod.104.030437.

Provvedini, D. M., C. D. Tsoukas, L. J. Deftos, and S. C. Manolagas. 1983. "1,25-dihydroxyvitamin D3 receptors in human leukocytes." *Science* 221 (4616):1181-3.

Quenby, S., and R. Farquharson. 2006. "Uterine natural killer cells, implantation failure and recurrent miscarriage." *Reprod Biomed Online* 13 (1):24-8.

Quraishi, S. A., G. De Pascale, J. S. Needleman, H. Nakazawa, M. Kaneki, E. K. Bajwa, C. A. Camargo, Jr., and I. Bhan. 2015. "Effect of Cholecalciferol Supplementation on Vitamin D Status and Cathelicidin Levels in Sepsis: A Randomized, Placebo-Controlled Trial." *Crit Care Med* 43 (9):1928-37. doi: 10.1097/CCM.0000000000001148.

Radman, M., A. Golshiri, A. Shamsizadeh, N. Zainodini, V. Bagheri, M. K. Arababadi, and D. Kennedy. 2015. "Toll-like receptor 4 plays significant roles during allergic rhinitis." *Allergol Immunopathol (Madr)* 43 (4):416-20. doi: 10.1016/j.aller.2014.04.006.

Raghupathy, R. 1997. "Th1-type immunity is incompatible with successful pregnancy." *Immunol Today* 18 (10):478-82.

Raghupathy, R., E. Al Mutawa, M. Makhseed, F. Azizieh, and J. Szekeres-Bartho. 2005. "Modulation of cytokine production by dydrogesterone in lymphocytes from women with recurrent miscarriage." *BJOG* 112 (8):1096-101. doi: 10.1111/j.1471-0528.2005.00633.x.

Rak, K., and M. Bronkowska. 2018. "Immunomodulatory Effect of Vitamin D and Its Potential Role in the Prevention and Treatment of Type 1 Diabetes Mellitus-A Narrative Review." *Molecules* 24 (1). doi: 10.3390/molecules24010053.

Ramos-Lopez, E., P. Bruck, T. Jansen, J. Herwig, and K. Badenhoop. 2007. "CYP2R1 (vitamin D 25-hydroxylase) gene is associated with susceptibility to type 1 diabetes and vitamin D levels in Germans." *Diabetes Metab Res Rev* 23 (8):631-6. doi: 10.1002/dmrr.719.

Ramos-Martinez, E., M. R. Lopez-Vancell, J. C. Fernandez de Cordova-Aguirre, J. Rojas-Serrano, A. Chavarria, A. Velasco-Medina, and G. Velazquez-Samano. 2018. "Reduction of respiratory infections in asthma patients supplemented with vitamin D is related to increased serum IL-10 and IFNgamma levels and cathelicidin expression." *Cytokine* 108:239-246. doi: 10.1016/j.cyto.2018.01.001.

Rigby, W. F., S. Denome, and M. W. Fanger. 1987. "Regulation of lymphokine production and human T lymphocyte activation by 1,25-dihydroxyvitamin D3. Specific inhibition at the level of messenger RNA." *J Clin Invest* 79 (6):1659-64. doi: 10.1172/JCI113004.

Rigby, W. F., T. Stacy, and M. W. Fanger. 1984. "Inhibition of T lymphocyte mitogenesis by 1,25-dihydroxyvitamin D3 (calcitriol)." *J Clin Invest* 74 (4):1451-5. doi: 10.1172/JCI111557.

Rosecrans, R., and J. C. Dohnal. 2014. "Seasonal vitamin D changes and the impact on health risk assessment." *Clin Biochem* 47 (7-8):670-2. doi: 10.1016/j.clinbiochem.2014.02.004.

Rubtsov, Y. P., J. P. Rasmussen, E. Y. Chi, J. Fontenot, L. Castelli, X. Ye, P. Treuting, L. Siewe, A. Roers, W. R. Henderson, Jr., W. Muller, and

A. Y. Rudensky. 2008. "Regulatory T cell-derived interleukin-10 limits inflammation at environmental interfaces." *Immunity* 28 (4):546-58. doi: 10.1016/j.immuni.2008.02.017.

Sadeghi, K., B. Wessner, U. Laggner, M. Ploder, D. Tamandl, J. Friedl, U. Zugel, A. Steinmeyer, A. Pollak, E. Roth, G. Boltz-Nitulescu, and A. Spittler. 2006. "Vitamin D3 down-regulates monocyte TLR expression and triggers hyporesponsiveness to pathogen-associated molecular patterns." *Eur J Immunol* 36 (2):361-70. doi: 10.1002/eji.200425995.

Sahin, O. A., D. Goksen, A. Ozpinar, M. Serdar, and H. Onay. 2017. "Association of vitamin D receptor polymorphisms and type 1 diabetes susceptibility in children: a meta-analysis." *Endocr Connect* 6 (3):159-171. doi: 10.1530/EC-16-0110.

Salazar Garcia, M. D., Y. Mobley, J. Henson, M. Davies, A. Skariah, S. Dambaeva, A. Gilman-Sachs, K. Beaman, C. Lampley, and J. Kwak-Kim. 2018. "Early pregnancy immune biomarkers in peripheral blood may predict preeclampsia." *J Reprod Immunol* 125:25-31. doi: 10.1016/j.jri.2017.10.048.

Sassi, F., C. Tamone, and P. D'Amelio. 2018. "Vitamin D: Nutrient, Hormone, and Immunomodulator." *Nutrients* 10 (11). doi: 10.3390/nu10111656.

Schaffler, H., D. P. Herlemann, P. Klinitzke, P. Berlin, B. Kreikemeyer, R. Jaster, and G. Lamprecht. 2018. "Vitamin D administration leads to a shift of the intestinal bacterial composition in Crohn's disease patients, but not in healthy controls." *J Dig Dis* 19 (4):225-234. doi: 10.1111/1751-2980.12591.

Scherberich, J. E., M. Kellermeyer, C. Ried, and A. Hartinger. 2005. "1-alpha-calcidol modulates major human monocyte antigens and toll-like receptors TLR 2 and TLR4 in vitro." *Eur J Med Res* 10 (4):179-82.

Shahrakyvahed, A., J. Sanchooli, N. Sanadgol, M. K. Arababadi, and D. Kennedy. 2014. "TLR9: an important molecule in the fight against hepatitis B virus." *Postgrad Med J* 90 (1065):396-401. doi: 10.1136/postgradmedj-2013-132309.

Shin, D. M., J. M. Yuk, H. M. Lee, S. H. Lee, J. W. Son, C. V. Harding, J. M. Kim, R. L. Modlin, and E. K. Jo. 2010. "Mycobacterial lipoprotein activates autophagy via TLR2/1/CD14 and a functional vitamin D receptor signalling." *Cell Microbiol* 12 (11):1648-65. doi: 10.1111/j.1462-5822.2010.01497.x.

Siklar, Z., D. Karatas, F. Dogu, B. Hacihamdioglu, A. Ikinciogullari, and M. Berberoglu. 2016. "Regulatory T Cells and Vitamin D Status in Children with Chronic Autoimmune Thyroiditis." *J Clin Res Pediatr Endocrinol* 8 (3):276-81. doi: 10.4274/jcrpe.2766.

Sly, L. M., M. Lopez, W. M. Nauseef, and N. E. Reiner. 2001. "1alpha,25-Dihydroxyvitamin D3-induced monocyte antimycobacterial activity is regulated by phosphatidylinositol 3-kinase and mediated by the NADPH-dependent phagocyte oxidase." *J Biol Chem* 276 (38):35482-93. doi: 10.1074/jbc.M102876200.

Song, G. G., S. C. Bae, and Y. H. Lee. 2012. "Association between vitamin D intake and the risk of rheumatoid arthritis: a meta-analysis." *Clin Rheumatol* 31 (12):1733-9. doi: 10.1007/s10067-012-2080-7.

Staeva-Vieira, T. P., and L. P. Freedman. 2002. "1,25-dihydroxyvitamin D3 inhibits IFN-gamma and IL-4 levels during in vitro polarization of primary murine CD4+ T cells." *J Immunol* 168 (3):1181-9.

Stene, L. C., G. Joner, and Group Norwegian Childhood Diabetes Study. 2003. "Use of cod liver oil during the first year of life is associated with lower risk of childhood-onset type 1 diabetes: a large, population-based, case-control study." *Am J Clin Nutr* 78 (6):1128-34. doi: 10.1093/ajcn/78.6.1128.

Stubbs, J. R., A. Idiculla, J. Slusser, R. Menard, and L. D. Quarles. 2010. "Cholecalciferol supplementation alters calcitriol-responsive monocyte proteins and decreases inflammatory cytokines in ESRD." *J Am Soc Nephrol* 21 (2):353-61. doi: 10.1681/ASN.2009040451.

Su, D., Y. Nie, A. Zhu, Z. Chen, P. Wu, L. Zhang, M. Luo, Q. Sun, L. Cai, Y. Lai, Z. Xiao, Z. Duan, S. Zheng, G. Wu, R. Hu, H. Tsukamoto, A. Lugea, Z. Liu, S. J. Pandol, and Y. P. Han. 2016. "Vitamin D Signaling through Induction of Paneth Cell Defensins Maintains Gut Microbiota and Improves Metabolic Disorders and Hepatic Steatosis in Animal Models." *Front Physiol* 7:498. doi: 10.3389/fphys.2016.00498.

Szekeres-Bartho, J. 2018. "The Role of Progesterone in Feto-Maternal Immunological Cross Talk." *Med Princ Pract* 27 (4):301-307. doi: 10.1159/000491576.

Szekeres-Bartho, J., F. Kilar, G. Falkay, V. Csernus, A. Torok, and A. S. Pacsa. 1985. "The mechanism of the inhibitory effect of progesterone on lymphocyte cytotoxicity: I. Progesterone-treated lymphocytes release a substance inhibiting cytotoxicity and prostaglandin synthesis." *Am J Reprod Immunol Microbiol* 9 (1):15-8.

Szekeres-Bartho, J., R. Kinsky, and G. Chaouat. 1990. "The effect of a progesterone-induced immunologic blocking factor on NK-mediated resorption." *Am J Reprod Immunol* 24 (4):105-7.

Szekeres-Bartho, J., and T. G. Wegmann. 1996. "A progesterone-dependent immunomodulatory protein alters the Th1/Th2 balance." *J Reprod Immunol* 31 (1-2):81-95.

Szilagyi, A., H. Leighton, B. Burstein, and X. Xue. 2014. "Latitude, sunshine, and human lactase phenotype distributions may contribute to geographic patterns of modern disease: the inflammatory bowel disease model." *Clin Epidemiol* 6:183-98. doi: 10.2147/CLEP.S59838.

Takeda, K., T. Kaisho, and S. Akira. 2003. "Toll-like receptors." *Annu Rev Immunol* 21:335-76. doi: 10.1146/annurev.immunol.21.120601.141126.

Tao, C., S. Simpson, Jr., I. van der Mei, L. Blizzard, E. Havrdova, D. Horakova, V. Shaygannejad, A. Lugaresi, G. Izquierdo, M. Trojano, P. Duquette, M. Girard, F. Grand'Maison, P. Grammond, R. Alroughani, M. Terzi, C. Oreja-Guevara, S. A. Sajedi, G. Iuliano, P. Sola, J. Lechner-Scott, V. V. Pesch, E. Pucci, R. Bergamaschi, M. Barnett, C. Ramo, B. Singhal, L. A. Spitaleri D, M. Slee, F. Verheul, R. Fernandez Bolanos, M. P. Amato, E. Cristiano, F. Granella, S. Hodgkinson, M. Fiol, O. Gray, P. McCombe, M. L. Saladino, J. L. Sanchez Menoyo, N. Shuey, S. Vucic, C. Shaw, N. Deri, W. O. Arruda, H. Butzkueven, T. Spelman, B. V. Taylor, and M. SBase Study Group. 2016. "Higher latitude is significantly associated with an earlier age of disease onset in multiple sclerosis." *J Neurol Neurosurg Psychiatry* 87 (12):1343-1349. doi: 10.1136/jnnp-2016-314013.

Thangamani, S., M. Kim, Y. Son, X. Huang, H. Kim, J. H. Lee, J. Cho, B. Ulrich, H. E. Broxmeyer, and C. H. Kim. 2015. "Cutting edge: progesterone directly upregulates vitamin d receptor gene expression for efficient regulation of T cells by calcitriol." *J Immunol* 194 (3):883-6. doi: 10.4049/jimmunol.1401923.

Thorsen, S. U., H. B. Mortensen, B. Carstensen, M. Fenger, B. H. Thuesen, L. Husemoen, R. Bergholdt, C. Brorsson, F. Pociot, A. Linneberg, and J. Svensson. 2014. "No association between type 1 diabetes and genetic variation in vitamin D metabolism genes: a Danish study." *Pediatr Diabetes* 15 (6):416-21. doi: 10.1111/pedi.12105.

Tian, Y., C. Wang, Z. Ye, X. Xiao, A. Kijlstra, and P. Yang. 2012. "Effect of 1,25-dihydroxyvitamin D3 on Th17 and Th1 response in patients with Behcet's disease." *Invest Ophthalmol Vis Sci* 53 (10):6434-41. doi: 10.1167/iovs.12-10398.

Tiosano, D., G. Wildbaum, V. Gepstein, O. Verbitsky, Y. Weisman, N. Karin, and A. Eztioni. 2013. "The role of vitamin D receptor in innate and adaptive immunity: a study in hereditary vitamin D-resistant rickets patients." *J Clin Endocrinol Metab* 98 (4):1685-93. doi: 10.1210/jc.2012-3858.

Tizaoui, K., W. Kaabachi, A. Hamzaoui, and K. Hamzaoui. 2014. "Contribution of VDR polymorphisms to type 1 diabetes susceptibility: Systematic review of case-control studies and meta-analysis." *J Steroid Biochem Mol Biol* 143:240-9. doi: 10.1016/j.jsbmb.2014.03.011.

Toldi, G., J. Rigo, Jr., B. Stenczer, B. Vasarhelyi, and A. Molvarec. 2011. "Increased prevalence of IL-17-producing peripheral blood lymphocytes in pre-eclampsia." *Am J Reprod Immunol* 66 (3):223-9. doi: 10.1111/j.1600-0897.2011.00987.x.

Toss, G., and T. Symreng. 1983. "Delayed hypersensitivity response and vitamin D deficiency." *Int J Vitam Nutr Res* 53 (1):27-31.

Treiber, G., B. Prietl, E. Frohlich-Reiterer, E. Lechner, A. Ribitsch, M. Fritsch, B. Rami-Merhar, C. Steigleder-Schweiger, W. Graninger, M. Borkenstein, and T. R. Pieber. 2015. "Cholecalciferol supplementation improves suppressive capacity of regulatory T-cells in young patients with new-onset type 1 diabetes mellitus - A randomized clinical trial." *Clin Immunol* 161 (2):217-24. doi: 10.1016/j.clim.2015.08.002.

Unger, W. W., S. Laban, F. S. Kleijwegt, A. R. van der Slik, and B. O. Roep. 2009. "Induction of Treg by monocyte-derived DC modulated by vitamin D3 or dexamethasone: differential role for PD-L1." *Eur J Immunol* 39 (11):3147-59. doi: 10.1002/eji.200839103.

van der Rhee, H. J., E. de Vries, and J. W. Coebergh. 2016. "Regular sun exposure benefits health." *Med Hypotheses* 97:34-37. doi: 10.1016/j.mehy.2016.10.011.

van Halteren, A. G., O. M. Tysma, E. van Etten, C. Mathieu, and B. O. Roep. 2004. "1alpha,25-dihydroxyvitamin D3 or analogue treated dendritic cells modulate human autoreactive T cells via the selective induction of apoptosis." *J Autoimmun* 23 (3):233-9. doi: 10.1016/j.jaut.2004.06.004.

Verma, R., and J. Y. Kim. 2016. "1,25-Dihydroxyvitamin D3 Facilitates M2 Polarization and Upregulates TLR10 Expression on Human Microglial Cells." *Neuroimmunomodulation* 23 (2):75-80. doi: 10.1159/000444300.

Wang, G., Q. Zhang, N. Xu, K. Xu, J. Wang, W. He, and T. Yang. 2014. "Associations between two polymorphisms (FokI and BsmI) of vitamin D receptor gene and type 1 diabetes mellitus in Asian population: a meta-analysis." *PLoS One* 9 (3):e89325. doi: 10.1371/journal.pone.0089325.

Wang, T. T., F. P. Nestel, V. Bourdeau, Y. Nagai, Q. Wang, J. Liao, L. Tavera-Mendoza, R. Lin, J. W. Hanrahan, S. Mader, and J. H. White. 2004. "Cutting edge: 1,25-dihydroxyvitamin D3 is a direct inducer of antimicrobial peptide gene expression." *J Immunol* 173 (5):2909-12.

Wei, R., and S. Christakos. 2015. "Mechanisms Underlying the Regulation of Innate and Adaptive Immunity by Vitamin D." *Nutrients* 7 (10):8251-60. doi: 10.3390/nu7105392.

White, J. H. 2010. "Vitamin D as an inducer of cathelicidin antimicrobial peptide expression: past, present and future." *J Steroid Biochem Mol Biol* 121 (1-2):234-8. doi: 10.1016/j.jsbmb.2010.03.034.

Wu, D., E. D. Lewis, M. Pae, and S. N. Meydani. 2018. "Nutritional Modulation of Immune Function: Analysis of Evidence, Mechanisms, and Clinical Relevance." *Front Immunol* 9:3160. doi: 10.3389/fimmu.2018.03160.

Xie, Z., J. Chen, C. Zheng, J. Wu, Y. Cheng, S. Zhu, C. Lin, Q. Cao, J. Zhu, and T. Jin. 2017. "1,25-dihydroxyvitamin D3 -induced dendritic cells suppress experimental autoimmune encephalomyelitis by increasing proportions of the regulatory lymphocytes and reducing T helper type 1 and type 17 cells." *Immunology* 152 (3):414-424. doi:10.1111/imm.12776.

Xu, H., A. Soruri, R. K. Gieseler, and J. H. Peters. 1993. "1,25-Dihydroxyvitamin D3 exerts opposing effects to IL-4 on MHC class-II antigen expression, accessory activity, and phagocytosis of human monocytes." *Scand J Immunol* 38 (6):535-40.

Zare-Bidaki, M., K. Tsukiyama-Kohara, and M. K. Arababadi. 2014. "Toll-like receptor 4 and hepatitis B infection: molecular mechanisms and pathogenesis." *Viral Immunol* 27 (7):321-6. doi:10.1089/vim.2014.0039.

Zhang, X., M. Zhou, Y. Guo, Z. Song, and B. Liu. 2015. "1,25-Dihydroxyvitamin D(3) Promotes High Glucose-Induced M1 Macrophage Switching to M2 via the VDR-PPARgamma Signaling Pathway." *Biomed Res Int* 2015:157834. doi: 10.1155/2015/157834.

Zhou, S. H., X. Wang, M. Y. Fan, H. L. Li, F. Bian, T. Huang, and H. Y. Fang. 2018. "Influence of vitamin D deficiency on T cell subsets and related indices during spinal tuberculosis." *Exp Ther Med* 16 (2):718-722. doi: 10.3892/etm.2018.6203.

Zipitis, C. S., and A. K. Akobeng. 2008. "Vitamin D supplementation in early childhood and risk of type 1 diabetes: a systematic review and meta-analysis." *Arch Dis Child* 93 (6):512-7. doi: 10.1136/adc.2007.128579.

In: Vitamin D as Progesterone-Like Hormone ISBN: 978-1-53617-516-5
Editor: Vittorio Unfer

Chapter 3

POSSIBLE ROLE OF VITAMIN D IN EMBRYO ADHESION AND IMPLANTATION

Maria Salomé Bezerra Espinola, MD
and Cesare Aragona*, MD
Systems Biology Group Lab,
Department of Experimental Medicine of
the "Sapienza" University of Rome, Rome, Italy

ABSTRACT

Vitamin D is a hormone with many features in common with progesterone. Evidence supports its positive role at the maternal-fetal interface. The demonstration that the injected oocyte and the embryo may remain for more than 120 hours in the uterine cavity, prior to implantation, strongly emphasizes the role of intrauterine nutrients before a direct contact between maternal blood and the blastocyst is established. In this regard vitamin D deserves special attention because of its possible role in the adhesion and the implantation processes in vitamin D deficient/insufficient women seeking pregnancy.

* Corresponding Author's Email: espinolasalome@gmail.com.

Keywords: vitamin D, progesterone-like hormone, embryo adhesion, implantation

Vitamin D Evolutionary Aspects

Vitamin D is an interesting hormone with a huge potential. This is probably linked to the fact that it has existed for more than 500-750 million years. Most animals and plants produce vitamin D. As other "old molecules" with such characteristic, it lost some activities along the evolutionary process (filogenesis/ontogenesis), when the unicellular organisms evolved into pluricellular structures and developed the hormonal and nervous systems to communicate. It has been more than 25 years since vitamin D has no longer been considered a hormone that regulates only calcium and bone health. Indeed, it is involved in many aspects in the areas of immunology, oncology, dermatology, rheumatology, hematology, endocrinology, obstetrics and reproductive medicine; about 2-3% of the human genes can be regulated by vitamin D (Holick 1989, 2013, Mazzaferro and Pasquali 2016).

VITAMIN (amine of life): an organic compound present in food and essential for the normal metabolism of organisms.

HORMONE (from Greek "ormao" = to start an action, to excite): an endogenous substance that acts both as a chemical messenger, transmitting signals from one cell to another, and as a transcriptional factor with genomic effects, capable of influencing the functions of the target organs.

Vitamin D Hormone with Progesterone-Like Activity

The structure of active vitamin D (calcitriol, 1,25$(OH)_2D_3$), revealed by the relevant studies of Holick and Norman (Holick, 1971; Norman, 1971), confirmed that it is actually a steroid hormone, and its receptor was "bona fide" accepted as member of the nuclear receptor gene family (Evans 1988, Mangelsdorf et al. 1995, Pike et al. 2016).

Table 1. Tissues that express the receptor for vitamin D (VDR)[1], Prolactin[2], Estradiol[3] and Progesterone[4]

Adipose [1,2,3]	Muscle, embryonic [1]
Adrenal [1,2,3]	Muscle, smooth [1]
Bone [1]	Osteoblast [1]
Bone marrow [1]	Ovary [1,2,3,4]
Brain [1,2,3,4]	Pancreas β cell [1]
Breast [1,2,3,4]	Parathyroid [1]
Cancer cells [1,2,4]	Parotid [1]
Cartilage [1]	Pituitary [1,2,3,4]
Colon [1]	Placenta [1,2]
Eggshell gland [1]	Prostate [1,2,3,4]
Epididymis [1,2]	Retina [1]
Hair follicle [1]	Skin [1]
Intestine [1]	Stomach [1]
Kidney [1,2]	Testis [1,2]
Liver (fetal) [1,2,3,4]	Thymus [1]
Lung [1]	Thyroid [1,2]
Lymphocytes (B & T) [1,2,4]	Uterus (endometrium, myometrium) [1,2,3,4]
Mucles, cardiac [1]	Yolk sac (bird) [1]

Many activities of $1,25(OH)_2D_3$ are mediated by the interaction with its nuclear receptor (Williams and Sigler 1998), leading to gene expression. But calcitriol can also exert its biological activity by binding to the plasma membrane and increasing the concentration of calcium inside the cell. Vitamin D receptor (VDR) was detected in over 36 human tissue/organs, e.g., B and T lymphocytes, hair follicles, muscles, adipose tissues, cancer cells and the bone marrow (Van Etten et al. 2003, Norman 2006, 2008). The plasma membrane VDR was discovered based on the rapid responses (rr) to $1,25(OH)_2D_3$, abnormally fast (minutes to an hour) to be explained as a consequence of the nuclear VDR regulating gene transcription (Haussler and Norman 1969, Tsai and Norman 1973, Norman, Mizwicki, and Norman

2004). Variations in either the hormone or the receptor concentration may lead to an altered response: physiological concentrations of both the hormone and the receptor determinate a normal response; low receptor concentrations and high hormone concentrations determinate a reduced or absent response; high receptor concentrations and low hormone concentrations determinate an enhanced response (Aragona and Friesen 1989). Tissue may express receptors for different hormones and the final biological effect represents the integrated response to all of them.

In particular, VDR is expressed in the ovary, the placenta and the uterus, suggesting a primary role in reproduction for vitamin D, as for estradiol and progesterone (Acker et al. 1982, Delvin et al. 1985, Kachkache et al. 1993, Kinuta et al. 2000, Vigano et al. 2006, Barrera et al. 2007, Parikh et al. 2010, Knight et al. 2010, Merhi et al. 2014, Dabrowski, Grzechocinska, and Wielgos 2015, Bakhshalizadeh et al. 2017, Cermisoni et al. 2018). Some similarities between progesterone and vitamin D are listed below (Monastra et al. 2018):

1. Chemical structure:
    a. Hydrofobic contacts.
    b. Hydrogen bond acceptor.
2. Hematic transport (globulin, transcortin). Only the free molecules can enter in the target cells.
3. Cytoplasmatic/Nuclear receptor: genomic effect. All the nuclear receptors of steroid hormones that show domains with common structure/function belong to the "nuclear receptor superfamily." Vitamin D possesses chemical features for binding to the progesterone receptor (Carlberg and Campbell 2013).
4. Membrane receptor: extragenomic effect.
5. Immunomodulatory activity: Vitamin D and Progesterone reciprocally enhance their effect.
6. Biological activities:
    a. Endometrial quality
    b. Blood supply
    c. Implantation window: adhesion and implantation.

The Preparation of the Uterus for Implantation

Decidualization of Endometrium

Early studies of vaginal cytology and endometrial histology, together with urinary steroid evaluation, allowed to divide the menstrual cycle in: 1) *follicular* or *proliferative phase* (pre-ovulation) (Li et al. 1991); 2) *ovulation*; 3) *secretory* or *luteal phase* (post-ovulation) (Zhu et al. 2014, Maybin and Critchley 2015). The dynamic relationships of the hypothalamic-pituitary-ovarian axis (gonadotropin-releasing hormone, follicle-stimulating hormone (FSH), luteinizing hormone (LH), granulosa cells, luteal cells, estrogen androgens, progesterone, inhibin) determine the recruitment, the growth, the selection, the maturation and the dominance of the ovulatory follicle giving rise to the corpus luteum. Subsequently, estradiol and progesterone, secreted by the corpus luteum, determine the proliferation and the fundamental secretory changes in the endometrial glands, whose shedding in the absence of fertilization and pregnancy determines the menstrual flow (Bhurke, Bagchi, and Bagchi 2016, Makieva et al. 2018). During the follicular phase of the menstrual cycle, estradiol concentrations gradually increase to reach a peak level 24 hours before ovulation, inducing the LH surge. Progesterone concentrations remain low during the whole follicular phase, and start rising only at the LH surge, as a direct effect of the FSH on granulosa cells. According to the two-cell, two-gonadotropin theory of estrogen biosynthesis, FSH stimulates granulosa cells to convert cholesterol to progesterone, which is transferred to the thecal cells and converted to androgens under the influence of LH. Androgens are then transferred to the granulosa cells and finally converted to estradiol. Such process is enhanced during the controlled ovarian hyperstimulation (COH) in Assisted Reproductive Technologies (ART), where multiple follicles grow and the follicular production of progesterone can be amplified. In the periovulatory phase, the concentration of estradiol falls rapidly to limit the duration of the LH surge. The formation of the corpus luteum on one hand shifts the steroidogenesis towards progesterone secretion, on the other hand promotes the structural and the secretory changes of the endometrial glands that favour the implantation (pre-decidualization and decidualization of the endometrium). The endometrial thickness is influenced by the estrogen levels, while blood supply is regulated by vitamin D and progesterone. Decidualization is the post ovulatory process of endometrial remodeling to prepare for pregnancy, and depends on cyclic adenosine monophosphate and progesterone signaling pathways, which influence both the transcriptome and the proteome. Decidualization encompasses secretory transformations of the uterine glands, vascular remodeling and influx of specialized uterine natural killer cells (NK) (Evans et al. 2016, Kim and Kim 2017, Ashary, Tiwari, and Modi 2018). Decidualizing cells can regulate trophoblast invasion, resist to inflammatory and oxidative processes and modulate local maternal immune responses under the influence of vitamin D and progesterone (Gellersen, Brosens, and Brosens 2007). First, glycogen-containing vacuoles appear between the nuclei and the basement membrane of the glandular cells of the endometrium. Then, secretion of nutrients increases, reaching a peak around 6 days after the ovulation. At this stage, the glands appear markedly tortuous. Pinopodes (greek for "drinking feet") are small protrusions that form on the endometrium between day 19 and 21 of the menstrual cycle and only persist for 2/3 days. Their development is enhanced by progesterone, while inhibited by estrogens. Receptive biomarkers associated with pinopodes are integrins, leukemia inhibiting factor (LIF), l-selectin, homeobox A10 (HOXA 10), glutaredoxin (GRLX), glycoledina (GLYA), heparin-binding epidermal growth factor (HB-EFG) and mucins. Pinopodes absorb, by endocytosis, most of the uterine fluids and the macromolecules during the early stage of implantation (Boron and Boulpaep 2005, Yang et al. 2017). As a consequence, the volume of the uterus decreases, bringing its walls closer to the embryoblast floating in it. Therefore, the period of active pinocytosis might limit the implantation window (Nikas and Psychoyos 1997, Nikas and Makrigiannakis 2003, Pantos et al. 2004, Rarani, Borhani, and Rashidi 2018, Paul, Sadek, and Mahesan 2019). Intrauterine microbiota also appears to play a major role in implantation, but its effects are still under intensive evaluation (Benner et al. 2018).

If the oocyte is fertilized, implantation occurs about eight days after ovulation, when endometrial stroma is loose and edematous.

In COH ART cycles, the administration of human chorionic gonadotropin (hCG) induces the luteal phase. As a result, the levels of progesterone rise to reach in 3-5 days the maximum values, necessary for the blastocyst transfer. Transfer in such time window also benefits from the routine admistration of progesterone either by vaginal, intra muscular or subcutaneous route (Table 2).

In summary, progesterone induces the decidualization of the endometrium and enhances the receptivity of the uterine cavity for the blastocyst transfer during the implantation window, relying on the secreted nutrients and the pinopode activity (Croxatto et al. 1978, Edwards, Steptoe, and Purdy 1980, Scott and Smith 1998, Pantos et al. 2004, Aragona et al. 2015, 2016).

The concentrations of some micronutrients (Zinc and Copper) change during the menstrual cycle, in consequence of the variation in ovarian and/or plasma volume (Michos et al. 2010). Similarly, Kim and co-workers observed changes in iron biomarkers such as hemoglobin, transferrin saturation, and ferritin (Kim, Yetley, and Calvo 1993). On the other hand, data on serum concentrations of vitamin D during the menstrual cycle failed to provide conclusive results, even though the concentrations of $1,25(OH)_2D_3$ undergo a 2/3-fold increase during pregnancy (Kumar et al. 1979). Regarding the circulating form of vitamin D, 25(OH)D, one study reports decreased concentrations in the peri-ovulatory and luteal phase, while four studies found no changes. Regarding $1,25(OH)_2D_3$, two studies found increased concentrations during the follicular phase, including a 128% increase from day 1 to 15 and a 56% increase from day 0 to 12. Two studies report increased concentrations from the follicular to the luteal phase by 13 and 26%, while five studies found no change in concentrations (Subramanian and Gernand 2019).

Table 2. Serum progesterone in 586 ART patients treated with various controlled ovarian hyperstimulation (COH) protocols. Means (±SD) are shown (Aragona, unpublished personal observations)

at hCG	0.8 (±0.3) ng/ml
at pick up	8.5 (±2.3) ng/ml
at embyro transfer (day 3-5)	27 (±3.2) ng/ml

Both the expression of vitamin D-metabolizing enzymes and the production of vitamin D at the maternal-fetal interface have been widely studied in pregnant humans and rodents. Furthermore vitamin D presence and concentration changes were observed in the plasma, the endometrial tissue and the uterine washing in pigs, supporting the role of vitamin D for the establishment and the maintenance of pregnancy in animal models and humans (Jang et al. 2017).

The Uterine Cavity Environment

Embryo implantation may occur outside the uterine cavity. In experimental animal models, indeed, implantation was observed also beneath the kidney capsule, in the testis capsule, in the anterior ocular chamber (Kirsby 1963). The fact that embryos implant outside the uterus but rarely inside the uterine cavity is only apparently a physiological paradox. Implantation is the "gateway of life" and, in order to create the best starting conditions, a perfect interaction between the accepting endometrium and the developing embryo is necessary; otherwise, implantation failure or pathological implantation occurs.

According to the fundamental work of Scott and Smith (Scott and Smith 1998), the uterine environment of most mammalian species studied can accommodate the embryo only in a narrow time span, called "implantation window." At this stage, the blastocyst stimulates site-specific changes (Psychoyos 1994) making a local part of the endometrium "receptive" for the implantion (Lopata 1996). Hertig et al. (Hertig, Rock, and Adams 1956) stated that the implantation window occurs from 6 days after the LH peak to 9 days after the ovulation (Xiao et al. 2010), or between 6 and 10 days post-ovulation (Elnashar and Aboul-Enein 2004). On average, it occurs between the 20th and the 23rd day after the last menstrual period.

Early work on Rhesus monkeys indicated that the primate uterus has a wider implantation window compared with other animals (Marston, Penn, and Sivelle 1977), thus a transfer of "day 2" embryos is possible also in humans (Edwards, Steptoe, and Purdy 1980, Edwards 2007b).

Inadequate uterine receptivity causes implantation failure in two-thirds of cases, while problems with the embryo itself accounts for the other third (Melford, Taylor, and Konje 2013). Abnormal cytokine and hormonal signaling, as well as epigenetic alterations, may be responsible for inadequate uterine receptivity and therefore recurrent implantation failure and female infertility. Thus, the optimization of endometrial receptivity can improve pregnancy rates. Furthermore, evaluating the markers of implantation may help to predict the pregnancy outcomes (Cakmak and Taylor 2011, Edgell, Rombauts, and Salamonsen 2013).

Luteal support consists in administrating medications, generally progestins, to increase the success rate of implantation and early embryogenesis, thereby complementing the function of the corpus luteum.

According to Grosser (Grosser 1910), preparing to implantation is the most important physiological function of the endometrial cycle. Under physiological conditions, human embryos generally reach the endometrial cavity about 96 hours after the LH surge (Croxatto et al. 1978). In artificial cycles, the fertilized embryos are transferred to the endometrial cavity one or two days earlier and, in general, successful pregnancy is achieved when the transfer occurs between days 17 and 19 (Li et al. 1991). Nowadays, the blastocyst transfer in natural or substituted cycles in "freeze all" programs appears to be the prevalent strategy used in routine clinical practice. Furthermore, term pregnancy and live birth following the immediate uterine transfer of sperm-microinjected oocytes in natural cycles has been reported by Aragona (Aragona et al. 2015, 2016).

Intra-Uterine Nourishment

As mentioned, the microinjected oocyte can be accepted, and develop up to the blastocyst stage, only during the implantation window, which relies on secreted nutrients and pinopode activity. Micro-injected oocytes proved to survive more than 120 hours in the uterine cavity, while embryoblasts approximately 72 hours before implanting. As they cannot receive nourishment directly from the blood of the mother during this time, oocytes or embryoblasts rely exclusively on nutrients secreted into the uterine cavity, e.g., iron, fat-soluble vitamins, several steroid-dependent proteins, cholesterol and steroids (Boron and Boulpaep 2005). Implantation is further facilitated by the synthesis of matrix substances, adhesion molecules and surface receptors for matrix substances. A list of such factors, some of which early mentioned by Seppälä (Seppala 2004), is reported in Table 3 (Hempstock et al. 2004).

Table 3. Factors identified in the uterine cavity

Matrix-associated	
	Fibronectin
	Laminin
	Entactin
	Type-IV collagen
	Heparan sulfate
	Proteoglycan
	Integrins
Others	
	Mucins
	Prolactin
	IGFBP-1
	Placental protein 5 (PP5)
	Placental protein 12 (PP12)
	Placental protein 14 (PP14) or glycodelin

	Pregnancy-associated endometrial alpha-1-globulin (alpha-1-PEG)
	Pregnancy-associated endometrial alpha-2-globulin (alpha-2-PEG)
	Endometrial protein 1-17
	17 β–hydroxysteroid dehydrogenase
	Albumin
	β–Lipoprotein
	Relaxin
	Fibroblast growth factor 1
	Fibroblast growth factor 2
	Pregnancy-associated plasma protein A (PAPP-A)
	Stress response protein 27 (SRP-27)
	CA-125
	β–Endorfin
	Leu-enkephalin
	Diamine oxidase
	Tissue plasminogen activator
	Renin
	Progesterone-dependent carbonic anhydrase
	Lactoferrin
	Alkaline phosphatase
	Progestagen-dependent endometrial protein
	Alpha uterine protein
	Chorionic alpha-2 microglobulin
	β–Lactoglobulin homologue
	Steroid hormones
	Vitamin D

The Role of Progesterone and Vitamin D at the Maternal-Fetal Interface

The embryo differs from the cells of the mother (Edwards 2007a), and would be rejected as a foreign organism by her immune system in the absence of secreted immunosuppressive agents. These include: platelet-activating factor, hCG, early pregnancy factor, immunosuppressive factor, prostaglandin E2, interleukin 1-alpha, interleukin 6, interferon-alpha, leukemia inhibitory factor and colony-stimulating factor (Whyte and Heap 1983, O'Neill et al. 1989, Mathialagan and Roberts 1994).

Factors released from the blastocyst also trigger the final formation of the decidual cells into their proper form. Some of them, in the proximity of the blastocyst, degenerate, thus providing nutrients for it (Turco et al. 2018).

Furthermore, hCG not only suppresses the immune response, but also signals the mother's body of the ongoing pregnancy (Edwards 2006) preventing menstruation by sustaining the function of the corpus luteum.

At this stage, the blastocyst secretes several factors, besides hCG, including: insulin-like growth factor 2; tumor necrosis factor-alpha; serine endopeptidase and metalloproteinases; a series of immunosuppressive agents such as platelet-activating factor, early pregnancy factor, immunosuppressive factor, prostaglandin E2, interleukin 1-alpha, interleukin 6, interferon-alpha, leukemia inhibitory factor, colony-stimulating factor; a series of others factors such as embryo-derived histamine releasing factor, tissue plasminogen activator and inhibitors, estradiol, beta 1-integrins, fibroblast growth factor, CYTL1, transforming growth factor alpha, inhibin preimplantation factor (Levin et al. 1984, Cocchiara et al. 1986, Ueyama et al. 2002, Ai, Jing, and Fang 2016, Kirkegaard et al. 2015).

Pregnancy presents a real challenge at the maternal-fetal interface, with fetal antigens potentially able to induce a reaction of the immune system of the mother. In this context, the bidirectional maternal-fetal interface communication guarantees the acceptance of the developing fetus and its protection from infections and inflammations (Sargent 1993, Szekeres-

Bartho 2002). Progesterone and Vitamin D (Tamblyn et al. 2015) seem to play a complex role by inducing antibacterial innate immune response to infection while suppressing adverse inflammatory adaptive immunity. Interestingly, the placenta is one of the principal sites for extra-renal synthesis of 1,25$(OH)_2D_3$. High decidua/placenta levels of vitamin D stimulate the immune function and promote the crosstalk between the maternal decidua and the fetal trophoblast. Decidual and placental specific levels of 1,25$(OH)_2D_3$ may be compromised in subjects with vitamin D insufficiency, as demonstrated both in humans and murine models. Proper acceptation of fetal antigens implies that the maternal immune system reacts with adaptive response mechanisms, some of which are indicated below (Najbauer, Szekeres-Bartho, and Tigyi 1989, Hewison 2012, Kim 2015, Subramanian and Gernand 2019):

1. Altered Th1/Th2 interplay
2. Appropriate humoral and cell mediated immunity
3. Progesterone dependent immunomodulatory response
4. Vitamin D-dependent immunomodulation. Increased expression of VDR by progesterone allows highly sensitive regulation of T cells by vitamin D, even when vitamin D levels are suboptimal. This mechanism effectively regulates T cells and prevents adverse immune responses during pregnancy (Thangamani et al. 2015).
5. Immunotolerance of the blastocyst, as an inducer of anti-inflammatory pathways (Th2 and Treg cells). In addition, progesterone and vitamin D stimulate the release of corticotropin-releasing factor (CRF), activin A, vascular endothelial growth factor (VEGF), osteopontin, calbindin. Moreover, both progesterone and vitamin D inhibit the pro-inflammatory pathways: NK, Th1 and Th17 cells (Faas and de Vos 2017). All these effects contribute in supporting pregnancy since its inception. It is noteworthy that their action is synergic (Monastra et al. 2018).
6. Progesterone-induced blocking factor (PIBF) production and immunomodulation.

Further indications in favor of the role of vitamin D at the maternal-fetal interface are reported below:

1. Serum level of vitamin D increases in pregnancy (Kumar 1980, Hartwell, Riis, and Christiansen 1990, Salle et al. 2000, Kumar and Yadav 2019).
2. Expression of VDR in the ovaries (steroidogenesis, aromatase), uterus, myometrium-endometrium-placenta (steroidogenesis).
3. VDR increases in pregnancy.
4. VDR expression in proliferative endometrial stromal cells, secretive endometrial stromal cells and pregnant decidual stroma cells (Vigano et al. 2006).
5. Vitamin D upregulates target genes as calbindin and osteopontin (implantation, decidualization) (Johnson et al. 2003).
6. Increased synthesis of vitamin D, induced by blastocyst-derived IL-1ß, determines Th1 reduction (Ota et al. 2015).
7. Immunomodulatory activity of vitamin D (Ota et al. 2015, Evans et al. 2006, Rafiee et al. 2015).
8. Progesterone enhances anti-inflammatory effects of vitamin D and viceversa.
9. Progesterone induces VDR in T-cells that counteract the immunological adverse reactions responsible for pregnancy complications (Thangamani et al. 2015).
10. Vitamin D reduces natural cytotoxicity.
11. Vitamin D reduces adherences of lymphocytes to antigens.
12. Vitamin D presents progesterone-like immunological effects.
13. Vitamin D induces PIBF production in activated human lymphocytes without competing for the progesterone receptors. Progesterone potentiates the effect of calcitriol probably via VDR induction.

It is important to stress that PIBFs represents a link between Progesterone and Vitamin D involved in the immune response during pregnancy (Raghupathy et al. 2009, Ermisch and Markert 2011, Bogdan,

Berta, and Szekeres-Bartho 2017, Szekeres-Bartho, Sucurovic, and Mulac-Jericevic 2018). PIBFs have anti-abortive effects, and inhibit lymphocytes cytotoxicity and prostaglandin secretion. PIBFs are released by T-gamma delta lymphocytes and decidual NK cells after stimulation with progesterone. Their immunomodulatory effect preserve pregnancy by:

1. Cytokine-like effect
2. Regulation of NK activity
3. Implantation and anti-abortive effect
4. Cell cycle regulation
5. Regulation of trophoblast invasiveness.

If all these delicate and complex mechanisms work, implantation of the embryo can finally occur passing through the following phases of the implantation window:

1. *Apposition*: a loose connection between the endometrium and the blastocyst likely mediated by MUC1, a glycoprotein expressed at the apical surface of endometrial epithelial cells, and by L selectin, expressed at the surface of the trophoblast.
2. *Adhesion*: a stronger communication between the endometrium and the blastocyst involving cytokines, chemokines in particular.
3. *Implantation*: the penetration of the trophoblast into the endometrium and its differentiation in syncytiotrophoblast. Basal membrane and stroma are invaded, contact with maternal blood is finally established.

VITAMIN D FROM EXPERIMENTAL TO CLINICAL EVIDENCE

According to the Italian Association of Clinical Endocrinologists (AME) and the Italian Chapter of American Association of Clinical Endocrinologists (AACE), values of serum vitamin D above 20 ng/mL (50 nmol/L) can be considered appropriate in the general population, with recommended levels above 30 ng/mL (75 nmol/L) in categories at risk: patients with diseases of bones, liver or kidneys; obese, pregnant or lactating subjects; patients with malabsorption conditions. In case of supplementation, dosages, vitamin D forms and modalities of administration can be individually tailored. No routine monitoring is recommended during vitamin D treatment, due to its large therapeutic index. In particular conditions, 25(OH)D levels can be measured after at least after a 6-month treatment (Norman et al. 2007, Cesareo et al. 2018, Carlberg 2019).

Vitamin D has been suggested to play a role in human reproduction (Luk et al. 2012, Lerchbaum and Obermayer-Pietsch 2012, Pagliardini et al. 2015), infertility and ART treatments (Garbedian et al. 2013, Vanni et al. 2014). Serum 25(OH)D levels may predict the ovarian response in women undergoing ovulation induction with clomiphene citrate (CC). Low levels of 25(OH)D and severe vitamin D deficiency (0,25 nmol/l or 0,10 ng/ml) were associated with lower rates of follicle development and pregnancy after ovarian stimulation with 50 mg CC (Ott et al. 2012). Adequate levels of vitamin D have also been suggested for reproductive success in women undergoing *in vitro* fertilization (IVF). In a prospective study on 84 women undergoing IVF, Ozkan et al. (Ozkan et al. 2010) assessed the 25(OH)D levels in the ovarian follicular fluid and determined an association between vitamin D status and cycle outcome. Follicular fluid levels of 25(OH)D were observed to be significantly higher in women achieving clinical pregnancy following fresh embryo transfer. Others failed to confirm these associations (Aleyasin et al. 2011). Furthermore, another group reported adverse effects of higher 25(OH)D levels on embryo quality (Anifandis et al. 2010), and basal vitamin D mean serum concentration (23.39 ± 6.60 ng/ml) failed to predict pregnancy in fertile controls or infertile patients (mean age 35.59 ± 4.47 years) admitted to our ART program (Aragona, unpublished personal observations). In line with the above considerations and the recommendations regarding serum vitamin D levels in pregnancy and lactation, integration is advisable in the case of deficient/insufficient women undergoing a pregnancy program.

CONCLUSION

Vitamin D, an old molecule involved in the evolutionary process, is currently considered a hormone, with many features in common with progesterone. While serum levels of progesterone are defined in the menstrual cycle, with a clear rise at the beginning of the luteal phase, inconclusive data on the serum levels of vitamin D have appeared in the literature. However, levels of VDR are present in human endometrial, decidual and placental tissues, and vitamin D levels were determined in pig intra-uterine fluid (Jang et al. 2017). On the other hand, progesterone can enhance the VDR activity, allowing vitamin D to exert its biological activities (regulation of endometrium adequacy and implantation window, immunomodulatory tolerance of the blastocyst) also at low concentrations at the maternal-fetal interface. The demonstration that the injected oocyte and the embryo can spend more than 120 hours in the uterine cavity before implanting (Aragona et al. 2015) strongly emphasizes the role of the intrauterine nutrients before a direct contact between maternal blood and the blastocyst is established. Nutraceutical are pharmaceutical-grade, standardized nutrients used as dietary supplements for preventing and treating diseases. Among them, vitamin D deserves special attention because of its possible role in the implantation process, especially in deficient/insufficient women undergoing a fertilization program. In this regard, *integration*, *implementation*, successful *implantation* represent a promising cascade sequence for vitamin D that still requires further evaluations with adequate clinical trials (Carlberg 2019).

REFERENCES

Acker, G. M., M. Garabedian, H. Guillozo, M. A. Pequinot, and S. Balsan. 1982. "*In Vitro* Formation of 25-Hydroxyvitamin D3 Metabolites in Endonietrium: Dependence on the Hormonal Status of the Rat*." *Endocrinology* 111 (6):2103-2109. doi: 10.1210/endo-111-6-2103.

Ai, Z., W. Jing, and L. Fang. 2016. "Cytokine-Like Protein 1(Cytl1): A Potential Molecular Mediator in Embryo Implantation." *PLoS One* 11 (1):e0147424. doi: 10.1371/journal.pone.0147424.

Aleyasin, A., M. A. Hosseini, A. Mahdavi, L. Safdarian, P. Fallahi, M. R. Mohajeri, M. Abbasi, and F. Esfahani. 2011. "Predictive value of the level of vitamin D in follicular fluid on the outcome of assisted reproductive technology." *Eur J Obstet Gynecol Reprod Biol* 159 (1):13 2-7. doi: 10.1016/j.ejogrb.2011.07.006.

Anifandis, Georgios M., Konstantinos Dafopoulos, Christina I. Messini, Nektarios Chalvatzas, Nikolaos Liakos, Spyros Pournaras, and Ioannis E. Messinis. 2010. "Prognostic value of follicular fluid 25-OH vitamin D and glucose levels in the IVF outcome." *Reproductive Biology and Endocrinology* 8 (1):91. doi: 10.1186/1477-7827-8-91.

Aragona, Cesare, and H G Friesen. 1989. "Lactation and galactorrhea." In *Endocrinology*, edited by L. J. DeGroot, 1613-1628. Saunders.

Aragona, Cesare, Antonella Linari, Giulietta Micara, Daniela Tranquilli, Agnese Maria Lourdes Gambaro, and Maria Salome Bezerra Espinola. 2015. "Term pregnancy and live birth subsequent to immediate uterine transfer of sperm microinjected oocyte in a natural cycle." *Gynecological Endocrinology* 31 (8):599-600. doi: 10.3109/09513590. 2015.1 031105.

Aragona, Cesare, Antonella Linari, Giulietta Micara, Daniela Tranquilli, Agnese Maria Lourdes Gambaro, and Maria Salome Bezerra Espinola. 2016. "Successful use of "Hour 1" transfer of sperm microinjected oocytes in natural cycles of poor responder women: a prospective randomized controlled pilot study." *Gynecological Endocrinology* 32 (5):370-373. doi: 10.3109/09513590.2015.1121227.

Ashary, Nancy, Abhishek Tiwari, and Deepak Modi. 2018. "Embryo Implantation: War in Times of Love." *Endocrinology* 159 (2):1188-1198. doi: 10.1210/en.2017-03082.

Bakhshalizadeh, S., F. Amidi, A. Alleyassin, M. Soleimani, R. Shirazi, and M. Shabani Nashtaei. 2017. "Modulation of steroidogenesis by vitamin D3 in granulosa cells of the mouse model of polycystic ovarian syndrome." *Syst Biol Reprod Med* 63 (3):150-161. doi: 10.1080/19396368.2017.1296046.

Barrera, D., E. Avila, G. Hernandez, A. Halhali, B. Biruete, F. Larrea, and L. Diaz. 2007. "Estradiol and progesterone synthesis in human placenta is stimulated by calcitriol." *J Steroid Biochem Mol Biol* 103 (3-5):529-32. doi: 10.1016/j.jsbmb.2006.12.097.

Benner, M., G. Ferwerda, I. Joosten, and R. G. van der Molen. 2018. "How uterine microbiota might be responsible for a receptive, fertile endometrium." *Hum Reprod Update* 24 (4):393-415. doi: 10.1093/ hum upd/dmy012.

Bhurke, Arpita S., Indrani C. Bagchi, and Milan K. Bagchi. 2016. "Progesterone-Regulated Endometrial Factors Controlling Implantation." *American journal of reproductive immunology (New York, N.Y.: 1989)* 75 (3):237-245. doi: 10.1111/aji.12473.

Bogdan, A., G. Berta, and J. Szekeres-Bartho. 2017. "PIBF positive uterine NK cells in the mouse decidua." *J Reprod Immunol* 119:38-43. doi: 10.1016/j.jri.2016.12.001.

Boron, W. F., and E. L. Boulpaep. 2005. *Medical Physiology: A Cellular and Molecular Approach*: Elsevier Saunders.

Cakmak, H., and H. S. Taylor. 2011. "Implantation failure: molecular mechanisms and clinical treatment." *Hum Reprod Update* 17 (2):242-53. doi: 10.1093/humupd/dmq037.

Carlberg, Carsten. 2019. "Nutrigenomics of Vitamin D." *Nutrients* 11 (3):676. doi: 10.3390/nu11030676.

Carlberg, Carsten, and Moray J. Campbell. 2013. "Vitamin D receptor signaling mechanisms: integrated actions of a well-defined transcription factor." *Steroids* 78 (2):127-136. doi: 10.1016/j.steroids. 2012.10.019.

Cermisoni, G. C., A. Alteri, L. Corti, E. Rabellotti, E. Papaleo, P. Vigano, and A. M. Sanchez. 2018. "Vitamin D and Endometrium: A Systematic Review of a Neglected Area of Research." *Int J Mol Sci* 19 (8). doi: 10.3390/ijms19082320.

Cesareo, R., R. Attanasio, M. Caputo, R. Castello, I. Chiodini, A. Falchetti, R. Guglielmi, E. Papini, A. Santonati, A. Scillitani, V. Toscano, V. Triggiani, F. Vescini, and M. Zini. 2018. "Italian Association of Clinical Endocrinologists (AME) and Italian Chapter of the American Association of Clinical Endocrinologists (AACE) Position Statement: Clinical Management of Vitamin D Deficiency in Adults." *Nutrients* 10 (5). doi: 10.3390/nu10050546.

Cocchiara, R., G. Di Trapani, A. Azzolina, G. Albeggiani, and D. Geraci. 1986. "Early embryonic histamine-releasing factor: a new model for human implantation." *Hum Reprod* 1 (7):445-7. doi: 10.1093/oxfordjournals.humrep.a136452.

Croxatto, H. B., M. E. Ortiz, S. Diaz, R. Hess, J. Balmaceda, and H. D. Croxatto. 1978. "Studies on the duration of egg transport by the human oviduct. II. Ovum location at various intervals following luteinizing hormone peak." *Am J Obstet Gynecol* 132 (6):629-34. doi: 10.1016/0002-9378(78)90854-2.

Dabrowski, F. A., B. Grzechocinska, and M. Wielgos. 2015. "The role of vitamin D in reproductive health--a Trojan Horse or the Golden Fleece?" *Nutrients* 7 (6):4139-53. doi: 10.3390/nu7064139.

Delvin, E. E., A. Arabian, F. H. Glorieux, and O. A. Mamer. 1985. "In vitro metabolism of 25-hydroxycholecalciferol by isolated cells from human decidua." *J Clin Endocrinol Metab* 60 (5):880-5. doi: 10.1210/jcem-60-5-880.

Edgell, T. A., L. J. Rombauts, and L. A. Salamonsen. 2013. "Assessing receptivity in the endometrium: the need for a rapid, non-invasive test." *Reprod Biomed Online* 27 (5):486-96. doi: 10.1016/j.rbmo.2013.05.014.

Edwards, R. G. 2006. "Genetics, epigenetics and gene silencing in differentiating mammalian embryos." *Reprod Biomed Online* 13 (5):732-53.

Edwards, R. G. 2007a. "Chorionic gonadotrophin, genes and embryonic signals regulating the implantation window." *Reprod Biomed Online* 14 (4):538-40.

Edwards, R. G. 2007b. "Human implantation: the last barrier in assisted reproduction technologies?" *Reprod Biomed Online* 14 Spec No 1:5-22. doi: 10.1016/s1472-6483(10)61454-x.

Edwards, R. G., P. C. Steptoe, and J. M. Purdy. 1980. "Establishing full-term human pregnancies using cleaving embryos grown in vitro." *Br J Obstet Gynaecol* 87 (9):737-56. doi: 10.1111/j.1471-0528.1980.tb0461 0.x.

Elnashar, A. M., and G. I Aboul-Enein. 2004. "Endometrial receptivity." *Middle East Fertility Society Journal* 9 (1):10-24.

Ermisch, C., and U. R. Markert. 2011. "[PIBF - Progesterone-Induced Blocking Factor]." *Z Geburtshilfe Neonatol* 215 (3):93-7. doi: 10.1055/ s-0031-1271742.

Evans, J., L. A. Salamonsen, A. Winship, E. Menkhorst, G. Nie, C. E. Gargett, and E. Dimitriadis. 2016. "Fertile ground: human endometrial programming and lessons in health and disease." *Nat Rev Endocrinol* 12 (11):654-667. doi: 10.1038/nrendo.2016.116.

Evans, Katie N., Lisa Nguyen, Junny Chan, Barbara A. Innes, Judith N. Bulmer, Mark D. Kilby, and Martin Hewison. 2006. "Effects of 25-Hydroxyvitamin D3 and 1,25-Dihydroxyvitamin D3 on Cytokine Production by Human Decidual Cells1." *Biology of Reproduction* 75 (6):816-822. doi: 10.1095/biolreprod.106.054056.

Evans, R. M. 1988. "The steroid and thyroid hormone receptor superfamily." *Science* 240 (4854):889-95. doi: 10.1126/science. 3283939.

Faas, M. M., and P. de Vos. 2017. "Uterine NK cells and macrophages in pregnancy." *Placenta* 56:44-52. doi: 10.1016/j.placenta.2017.03.001.

Garbedian, Kimberley, Miranda Boggild, Joel Moody, and Kimberly E. Liu. 2013. Effect of vitamin D status on clinical pregnancy rates following in vitro fertilization. *CMAJ open* 1 (2): E77-82. Accessed 2013/05//. doi:10.9778/cmajo.20120032.

Gellersen, B., I. A. Brosens, and J. J. Brosens. 2007. "Decidualization of the human endometrium: mechanisms, functions, and clinical perspectives." *Semin Reprod Med* 25 (6):445-53. doi: 10.1055/s-2007-991042.

Grosser, Otto. 1910. "The development of the egg membranes and the placenta; menstruation." In *Manual of human embryology*, edited by Franz Keibel and Franklin P. Mall, 91-179. philadelphia: J. B. Lippincott Company.

Hartwell, D., B. J. Riis, and C. Christiansen. 1990. "Changes in vitamin D metabolism during natural and medical menopause." *J Clin Endocrinol Metab* 71 (1):127-32. doi: 10.1210/jcem-71-1-127.

Haussler, M. R., and A. W. Norman. 1969. "Chromosomal receptor for a vitamin D metabolite." *Proceedings of the National Academy of Sciences of the United States of America* 62 (1):155-162. doi: 10.1073/pnas.62.1.155.

Hempstock, J., T. Cindrova-Davies, E. Jauniaux, and G. J. Burton. 2004. "Endometrial glands as a source of nutrients, growth factors and cytokines during the first trimester of human pregnancy: a morphological and immunohistochemical study." *Reprod Biol Endocrinol* 2:58. doi: 10.1186/1477-7827-2-58.

Hertig, A. T., J. Rock, and E. C. Adams. 1956. "A description of 34 human ova within the first 17 days of development." *Am J Anat* 98 (3):435-93.

Hewison, M. 2012. "An update on vitamin D and human immunity." *Clin Endocrinol (Oxf)* 76 (3):315-25. doi: 10.1111/j.1365-2265.2011.0426 1.x.

Holick, M. F., Schnoes H. K., DeLuca H. F. 1971. "Identification of 1,25-dihydroxycholecalciferol, a form of vitamin D3 metabolically active in the intestine." Proc Natl Acad Sci U S A. 68(4):803-4.

Holick, M F. 1989. "Vitamin D: Biosynthesis, metabolism and Mode of Action." In *Endocrinology*, edited by L. J. DeGroot, 902-926. Saunders.

Holick, M F. 2013. "Evolution and Function of Vitamin D " In *Vitamin D Analogs in Cancer Prevention and Therapy*, edited by J. Reichrath, M. Friedrich and W. Tilgen, 3-28. Springer Berlin Heidelberg.

Jang, Hwanhee, Yohan Choi, Inkyu Yoo, Jisoo Han, Jin Su Hong, Yoo Yong Kim, and Hakhyun Ka. 2017. "Vitamin D-metabolic enzymes and related molecules: Expression at the maternal-conceptus interface and the role of vitamin D in endometrial gene expression in pigs." *PloS one* 12 (10):e0187221-e0187221. doi: 10.1371/journal.pone.0187221.

Johnson, G. A., R. C. Burghardt, F. W. Bazer, and T. E. Spencer. 2003. "Osteopontin: roles in implantation and placentation." *Biol Reprod* 69 (5):1458-71. doi: 10.1095/biolreprod.103.020651.

Kachkache, M., C. Rebut-Bonneton, J. Demignon, E. Cynober, and M. Garabedian. 1993. "Uterine cells other than stromal decidual cells are required for 1,25-dihydroxyvitamin D3 production during early human pregnancy." *FEBS Lett* 333 (1-2):83-8. doi: 10.1016/0014-5793(93) 80379-9.

Kim, C. H. 2015. "A functional relay from progesterone to vitamin D in the immune system." *DNA Cell Biol* 34 (6):379-82. doi: 10.1089/ dna.2015.2857.

Kim, I., E. A. Yetley, and M. S. Calvo. 1993. "Variations in iron-status measures during the menstrual cycle." *Am J Clin Nutr* 58 (5):705-9. doi: 10.1093/ajcn/58.5.705.

Kim, Su-Mi, and Jong-Soo Kim. 2017. "A Review of Mechanisms of Implantation." *Development & reproduction* 21 (4):351-359. doi: 10.12717/DR.2017.21.4.351.

Kinuta, K., H. Tanaka, T. Moriwake, K. Aya, S. Kato, and Y. Seino. 2000. "Vitamin D is an important factor in estrogen biosynthesis of both female and male gonads." *Endocrinology* 141 (4):1317-24. doi: 10.1210/endo.141.4.7403.

Kirkegaard, K., P. Villesen, J. M. Jensen, J. J. Hindkjaer, S. Kolvraa, H. J. Ingerslev, and K. Lykke-Hartmann. 2015. "Distinct differences in global gene expression profiles in non-implanted blastocysts and blastocysts resulting in live birth." *Gene* 571 (2):212-20. doi: 10.1016/j. gene.2015.06.057.

Kirsby, D. R. S. K. 1963. "The development of mouse blastocysts transplanted to the scrotal and cryptorchid testis." *J. Anal., Lond.* 97, 1, pp. 119-130.

Knight, J. A., J. Wong, K. M. Blackmore, J. M. Raboud, and R. Vieth. 2010. "Vitamin D association with estradiol and progesterone in young women." *Cancer Causes Control* 21 (3):479-83. doi: 10.1007/s10552-009-9466-0.

Kumar, J., and A. Yadav. 2019. "Vitamin D deficiency pandemic among pregnant women." *J Family Med Prim Care* 8 (4):1515-1516. doi: 10.4103/jfmpc.jfmpc_202_19.

Kumar, R. 1980. "The metabolism of 1,25-dihydroxyvitamin D3." *Endocr Rev* 1 (3):258-67. doi: 10.1210/edrv-1-3-258.

Kumar, R., W. R. Cohen, P. Silva, and F. H. Epstein. 1979. "Elevated 1,25-dihydroxyvitamin D plasma levels in normal human pregnancy and lactation." *The Journal of clinical investigation* 63 (2):342-344. doi: 10.1172/JCI109308.

Lerchbaum, E., and B. Obermayer-Pietsch. 2012. "Vitamin D and fertility: a systematic review." *Eur J Endocrinol* 166 (5):765-78. doi: 10.1530/eje-11-0984.

Levin, E. G., U. Marzec, J. Anderson, and L. A. Harker. 1984. "Thrombin stimulates tissue plasminogen activator release from cultured human endothelial cells." *J Clin Invest* 74 (6):1988-95. doi: 10.1172/ jci111620.

Li, T. C., M. A. Warren, P. Dockery, and I. D. Cooke. 1991. "Human endometrial morphology around the time of implantation in natural and artificial cycles." *J Reprod Fertil* 92 (2):543-54. doi: 10.1530/jrf.0. 0920543.

Lopata, A. 1996. "Blastocyst-endometrial interaction: an appraisal of some old and new ideas." *Mol Hum Reprod* 2 (7):519-25. doi: 10.1093/ molehr/2.7.519.

Luk, J., S. Torrealday, G. Neal Perry, and L. Pal. 2012. "Relevance of vitamin D in reproduction." *Hum Reprod* 27 (10):3015-27. doi: 10.1093/humrep/des248.

Makieva, S., E. Giacomini, J. Ottolina, A. M. Sanchez, E. Papaleo, and P. Vigano. 2018. "Inside the Endometrial Cell Signaling Subway: Mind the Gap(s)." *Int J Mol Sci* 19 (9). doi: 10.3390/ijms19092477.

Mangelsdorf, D. J., C. Thummel, M. Beato, P. Herrlich, G. Schutz, K. Umesono, B. Blumberg, P. Kastner, M. Mark, P. Chambon, and R. M. Evans. 1995. "The nuclear receptor superfamily: the second decade." *Cell* 83 (6):835-9. doi: 10.1016/0092-8674(95)90199-x.

Marston, J. H., R. Penn, and P. C. Sivelle. 1977. "Successful autotransfer of tubal eggs in the rhesus monkey (Macaca mulatta)." *J Reprod Fertil* 49 (1):175-6. doi: 10.1530/jrf.0.0490175.

Mathialagan, N., and R. M. Roberts. 1994. "A role for cytokines in early pregnancy." *Indian J Physiol Pharmacol* 38 (3):153-62.

Maybin, J. A., and H. O. Critchley. 2015. "Menstrual physiology: implications for endometrial pathology and beyond." *Hum Reprod Update* 21 (6):748-61. doi: 10.1093/humupd/dmv038.

Mazzaferro, S., and M. Pasquali. 2016. "Vitamin D: a dynamic molecule. How relevant might the dynamism for a vitamin be?" *Nephrol Dial Transplant* 31 (1):23-30. doi: 10.1093/ndt/gfv340.

Melford, Sarah E., Anthony H. Taylor, and Justin C. Konje. 2013. "Of mice and (wo)men: factors influencing successful implantation including endocannabinoids." *Human Reproduction Update* 20 (3):415-428. doi: 10.1093/humupd/dmt060.

Merhi, Z., A. Doswell, K. Krebs, and M. Cipolla. 2014. "Vitamin D alters genes involved in follicular development and steroidogenesis in human cumulus granulosa cells." *J Clin Endocrinol Metab* 99 (6):E1137-45. doi: 10.1210/jc.2013-4161.

Michos, C., V. Kalfakakou, S. Karkabounas, D. Kiortsis, and A. Evangelou. 2010. "Changes in copper and zinc plasma concentrations during the normal menstrual cycle in women." *Gynecol Endocrinol* 26 (4):250-5. doi: 10.3109/09513590903247857.

Monastra, G., S. De Grazia, L. De Luca, S. Vittorio, and V. Unfer. 2018. "Vitamin D: a steroid hormone with progesterone-like activity." *Eur Rev Med Pharmacol Sci* 22 (8):2502-2512. doi: 10.26355/eurrev_201804_14845.

Najbauer, J., J. Szekeres-Bartho, and G. J. Tigyi. 1989. "Modulation of cell-cell and cell-antigen interactions by 1,25-dihydroxyvitamin D3 and vitamin D3 sulfate in vitro: a study on pregnancy lymphocytes and hybridoma cells." *Immunol Lett* 20 (4):317-22. doi: 10.1016/0165-2478(89)90041-2.

Nikas, G., and A. Makrigiannakis. 2003. "Endometrial pinopodes and uterine receptivity." *Ann N Y Acad Sci* 997:120-3. doi: 10.1196/annals. 1290.042.

Nikas, G., and A. Psychoyos. 1997. "Uterine pinopodes in peri-implantation human endometrium. Clinical relevance." *Ann N Y Acad Sci* 816:129-42. doi: 10.1111/j.1749-6632.1997.tb52136.x.

Norman, A. W., Myrtle, J. F., Midgett, R. J., Nowicki, H. G., Williams, V., Popjak, G. 1971. "1,25-dihydroxycholecalciferol: identification of the proposed active form of vitamin D3 in the intestine." *Science* 17351-54:51-54.

Norman, A. W. 2006. "Minireview: vitamin D receptor: new assignments for an already busy receptor." *Endocrinology* 147 (12):5542-8. doi: 10.1210/en.2006-0946.

Norman, A. W. 2008. "From vitamin D to hormone D: fundamentals of the vitamin D endocrine system essential for good health." *Am J Clin Nutr* 88 (2):491s-499s. doi: 10.1093/ajcn/88.2.491S.

Norman, A. W., M. T. Mizwicki, and D. P. Norman. 2004. "Steroid-hormone rapid actions, membrane receptors and a conformational ensemble model." *Nat Rev Drug Discov* 3 (1):27-41. doi: 10.1038/ nrd1283.

Norman, Anthony W., Roger Bouillon, Susan J. Whiting, Reinhold Vieth, and Paul Lips. 2007. "13th Workshop consensus for vitamin D nutritional guidelines." *The Journal of steroid biochemistry and molecular biology* 103 (3-5):204-205. doi:10.1016/j.jsbmb.2006.12.07 1.

O'Neill, C., J. P. Ryan, M. Collier, D. M. Saunders, A. J. Ammit, and I. L. Pike. 1989. "Supplementation of in-vitro fertilisation culture medium with platelet activating factor." *Lancet* 2 (8666):769-72. doi:10.1016/s 0140-6736(89)90831-3.

Ota, Kuniaki, Svetlana Dambaeva, Michael Woo-Il Kim, Ae-Ra Han, Atsushi Fukui, Alice Gilman-Sachs, Kenneth Beaman, and Joanne Kwak-Kim. 2015. "1,25-Dihydroxy-vitamin D3 regulates NK-cell cytotoxicity, cytokine secretion, and degranulation in women with recurrent pregnancy losses." *European Journal of Immunology* 45 (11):3188-3199. doi: 10.1002/eji.201545541.

Ott, J., L. Wattar, C. Kurz, R. Seemann, J. C. Huber, K. Mayerhofer, and E. Vytiska-Binstorfer. 2012. "Parameters for calcium metabolism in women with polycystic ovary syndrome who undergo clomiphene citrate stimulation: a prospective cohort study." *Eur J Endocrinol* 166 (5):897-902. doi: 10.1530/eje-11-1070.

Ozkan, S., S. Jindal, K. Greenseid, J. Shu, G. Zeitlian, C. Hickmon, and L. Pal. 2010. "Replete vitamin D stores predict reproductive success following in vitro fertilization." *Fertil Steril* 94 (4):1314-9. doi: 10.1016/j.fertnstert.2009.05.019.

Pagliardini, L., P. Vigano, M. Molgora, P. Persico, A. Salonia, S. H. Vailati, A. Paffoni, E. Somigliana, E. Papaleo, and M. Candiani. 2015. "High Prevalence of Vitamin D Deficiency in Infertile Women Referring for Assisted Reproduction." *Nutrients* 7 (12):9972-84. doi:10.3390/nu71 25516.

Pantos, K., G. Nikas, E. Makrakis, D. Stavrou, P. Karantzis, and M. Grammatis. 2004. "Clinical value of endometrial pinopodes detection in artificial donation cycles." *Reprod Biomed Online* 9 (1):86-90.

Parikh, G., M. Varadinova, P. Suwandhi, T. Araki, Z. Rosenwaks, L. Poretsky, and D. Seto-Young. 2010. "Vitamin D regulates steroidogenesis and insulin-like growth factor binding protein-1 (IGFBP-1) production in human ovarian cells." *Horm Metab Res* 42 (10):754-7. doi: 10.1055/s-0030-1262837.

Paul, A. B. M., S. T. Sadek, and A. M. Mahesan. 2019. "The role of microRNAs in human embryo implantation: a review." *J Assist Reprod Genet* 36 (2):179-187. doi: 10.1007/s10815-018-1326-y.

Pike, J. Wesley, Mark B. Meyer, Nancy A. Benkusky, Seong Min Lee, Hillary St John, Alex Carlson, Melda Onal, and Sohel Shamsuzzaman. 2016. "Genomic Determinants of Vitamin D-Regulated Gene Expression." *Vitamins and hormones* 100:21-44. doi: 10.1016/bs.vh. 2015.10.011.

Psychoyos, A. 1994. "The implantation window: basic and clinical aspects." In *Perspectives on Assisted Reproduction*, edited by T. Mori, 57-62. Ares-Serono symposia publications.

Rafiee, M., M. Gharagozloo, A. Ghahiri, F. Mehrabian, M. R. Maracy, S. Kouhpayeh, I. L. Pieper, and A. Rezaei. 2015. "Altered Th17/Treg Ratio in Recurrent Miscarriage after Treatment with Paternal Lymphocytes and Vitamin D3: a Double-Blind Placebo-Controlled Study." *Iran J Immunol* 12 (4):252-62. doi: IJIv12i4A3.

Raghupathy, R., E. Al-Mutawa, M. Al-Azemi, M. Makhseed, F. Azizieh, and J. Szekeres-Bartho. 2009. "Progesterone-induced blocking factor (PIBF) modulates cytokine production by lymphocytes from women with recurrent miscarriage or preterm delivery." *J Reprod Immunol* 80 (1-2):91-9. doi: 10.1016/j.jri.2009.01.004.

Rarani, F. Z., F. Borhani, and B. Rashidi. 2018. "Endometrial pinopode biomarkers: Molecules and microRNAs." *J Cell Physiol* 233 (12):9145-9158. doi: 10.1002/jcp.26852.

Salle, Bernard L, Edgar E Delvin, Alexandre Lapillonne, Nicholas J Bishop, and Francis H Glorieux. 2000. "Perinatal metabolism of vitamin D." *The American Journal of Clinical Nutrition* 71 (5):1317S-1324S. doi: 10.1093/ajcn/71.5.1317s.

Sargent, I. L. 1993. "Maternal and fetal immune responses during pregnancy." *Exp Clin Immunogenet* 10 (2):85-102.

Scott, L. A., and S. Smith. 1998. "The successful use of pronuclear embryo transfers the day following oocyte retrieval." *Hum Reprod* 13 (4):1003-13. doi: 10.1093/humrep/13.4.1003.

Seppala, M. 2004. "Advances in uterine protein research: reproduction and cancer." *Int J Gynaecol Obstet* 85 (2):105-18. doi: 10.1016/j.ijgo.2004.01.007.

Subramanian, Anita, and Alison D. Gernand. 2019. "Vitamin D metabolites across the menstrual cycle: a systematic review." *BMC Women's Health* 19 (1):19. doi: 10.1186/s12905-019-0721-6.

Szekeres-Bartho, J. 2002. "Immunological relationship between the mother and the fetus." *Int Rev Immunol* 21 (6):471-95.

Szekeres-Bartho, J., S. Sucurovic, and B. Mulac-Jericevic. 2018. "The Role of Extracellular Vesicles and PIBF in Embryo-Maternal Immune-Interactions." *Front Immunol* 9:2890. doi: 10.3389/fimmu.2018.02890.

Tamblyn, J A, M Hewison, C L Wagner, J N Bulmer, and M D Kilby. 2015. "Immunological role of vitamin D at the maternal–fetal interface." 224 (3):R107. doi: 10.1530/joe-14-0642.

Thangamani, S., M. Kim, Y. Son, X. Huang, H. Kim, J. H. Lee, J. Cho, B. Ulrich, H. E. Broxmeyer, and C. H. Kim. 2015. "Cutting edge: progesterone directly upregulates vitamin d receptor gene expression for efficient regulation of T cells by calcitriol." *J Immunol* 194 (3):883-6. doi: 10.4049/jimmunol.1401923.

Tsai, Huan C., and Anthony W. Norman. 1973. "Studies on the mode of action of calciferol VI: Effect of 1,25-dihydroxy-vitamin D3 on RNA synthesis in the intestinal mucosa." *Biochemical and Biophysical Research Communications* 54 (2):622-627. doi: doi.org/10.1016/0006-291X(73)91468-X.

Turco, M. Y., L. Gardner, R. G. Kay, R. S. Hamilton, M. Prater, M. S. Hollinshead, A. McWhinnie, L. Esposito, R. Fernando, H. Skelton, F. Reimann, F. M. Gribble, A. Sharkey, S. G. E. Marsh, S. O'Rahilly, M. Hemberger, G. J. Burton, and A. Moffett. 2018. "Trophoblast organoids as a model for maternal-fetal interactions during human placentation." *Nature* 564 (7735):263-267. doi: 10.1038/s41586-018-0753-3.

Ueyama, M., N. Kasatori, T. Urayama, T. Maemura, Y. Yao, T. Shiraishi, S. Saito, and H. Kubo. 2002. "Quantitative evaluation of the influence of ovarian steroids on plasminogen activators and inhibitors in human endometrial cells and trophoblasts." *Thromb Res* 108 (4):235-44. doi: 10.1016/s0049-3848(03)00029-x.

Van Etten, Evelyne, Brigitte Decallonne, Lieve Verlinden, Annemieke Verstuyf, Roger Bouillon, and Chantal Mathieu. 2003. "Analogs of 1α,25-dihydroxyvitamin D3 as pluripotent immunomodulators." *Journal of Cellular Biochemistry* 88 (2):223-226. doi: 10.1002/jcb.10329.

Vanni, V. S., P. Vigano, E. Somigliana, E. Papaleo, A. Paffoni, L. Pagliardini, and M. Candiani. 2014. "Vitamin D and assisted reproduction technologies: current concepts." *Reprod Biol Endocrinol* 12:47. doi: 10.1186/1477-7827-12-47.

Vigano, P., D. Lattuada, S. Mangioni, L. Ermellino, M. Vignali, E. Caporizzo, P. Panina-Bordignon, M. Besozzi, and A. M. Di Blasio. 2006. "Cycling and early pregnant endometrium as a site of regulated expression of the vitamin D system." *J Mol Endocrinol* 36 (3):415-24. doi: 10.1677/jme.1.01946.

Whyte, A., and R. B. Heap. 1983. "Reproductive immunology. Early pregnancy factor." *Nature* 304 (5922):121-2. doi: 10.1038/304121a0.

Williams, S. P., and P. B. Sigler. 1998. "Atomic structure of progesterone complexed with its receptor." *Nature* 393 (6683):392-6. doi: 10.1038/30775.

Xiao, Y., X. Sun, X. Yang, J. Zhang, Q. Xue, B. Cai, and Y. Zhou. 2010. "Leukemia inhibitory factor is dysregulated in the endometrium and uterine flushing fluid of patients with adenomyosis during implantation window." *Fertil Steril* 94 (1):85-9. doi: 10.1016/j.fertnstert.2009.03.012.

Yang, Y., X. Chen, S. H. Saravelos, Y. Liu, J. Huang, J. Zhang, and T. C. Li. 2017. "HOXA-10 and E-cadherin expression in the endometrium of women with recurrent implantation failure and recurrent miscarriage." *Fertil Steril* 107 (1):136-143.e2. doi: 10.1016/j.fertnstert.2016.09.016.

Zhu, H., C. C. Hou, L. F. Luo, Y. J. Hu, and W. X. Yang. 2014. "Endometrial stromal cells and decidualized stromal cells: origins, transformation and functions." *Gene* 551 (1):1-14. doi: 10.1016/j. gene.2014.08.047.

In: Vitamin D as Progesterone-Like Hormone ISBN: 978-1-53617-516-5
Editor: Vittorio Unfer

Chapter 4

Vitamin D for the Prevention of Miscarriage and Spontaneous Preterm Birth

Daniela Menichini[1,2] *, MD and Fabio Facchinetti [2], MD

[1] Department of Biomedical, Metabolic and Neural Sciences, International Doctorate School in Clinical and Experimental Medicine, University of Modena and Reggio Emilia, Modena, Italy
[2] Mother Infant Department, University of Modena and Reggio Emilia, Modena, Italy

Abstract

Pregnancy represents a dynamic period with physical and physiological changes in both the mother and the developing fetus. Vitamin D was shown to exert many physiological activities during the very early stages of gestation in perfect synchrony with progesterone. Both the molecules mutually help and reinforce the activity exerted by each one. Vitamin D contributes to prepare the endometrium to be receptive and

* Corresponding Author's Email: daniela.menichini91@gmail.com.

supports the implantation process and the course of pregnancy through similar pathways to those used by progesterone, giving rise to a significant synergy of action. It is increasingly evident that vitamin D gives an essential support from the luteal phase onwards exerting an important role in the establishment and maintenance of pregnancy through its immunomodulatory action and its effects on inflammation within the placenta, preventing miscarriage and preterm birth (PTB). Thus, it is important to guarantee the appropriate levels of vitamin D before placentation (and trophoblast invasion); however further research is needed to decipher vitamin D requirements of pregnant women and the optimal timing of supplementation, taking into account a variety of lifestyles, body types and baseline vitamin D status. Determining the role of vitamin D in non-classical, immune pathways continues to be a challenge that once answered will substantiate recommendations and public health policies.

Keywords: vitamin D, miscarriage, recurrent pregnancy loss, preterm birth

INTRODUCTION

Vitamin D, a fat-soluble secosteroid hormone, is synthesized primarily in the skin upon exposure to sunlight and is converted to active vitamin D3 in the liver and the kidney. The classical function of vitamin D is regulating calcium and phosphorus homeostasis (Wacker and Holick 2013). Recently, vitamin D is well recognized for its non-classical actions including antioxidant activity, modulation of innate immune response and adaptive immune response, bone mineralization, cellular growth, and has been shown to decrease the risk for chronic illnesses such as diabetes, cardiovascular disease, cancer, obesity, and autoimmune diseases (Chen et al. 2018). Vitamin D receptors (VDR) allow the biological activity of vitamin D and are found in many cells throughout the body. VDR have been identified in reproductive tissues such as human testis, sperm, epididymis, seminal vesicle, prostate, ovaries, uterus, placenta, cervix, breast tissue, the pituitary, and hypothalamus (Irani and Merhi 2014) (Figure 1). Vitamin D has recently received attention for the role it plays in reproduction and fertility (Sollis 2015). Animal models showed that vitamin D deficiency is associated with impaired ovarian function and decreased fertility (Halloran and Deluca

1980). Human studies suggest that vitamin D concentrations, in both serum and follicular fluid, are predictive of *in vitro* fertilisation (IVF) success via effects on endometrial receptivity and implantation (Ozkan et al. 2010). Furthermore, because of its role as an immune modulator, vitamin D is thought to prevent pregnancy loss by facilitating maternal immune tolerance during pregnancy. Through these mechanisms, vitamin D might influence fecundity, livebirth, and pregnancy loss (Mumford et al. 2018).

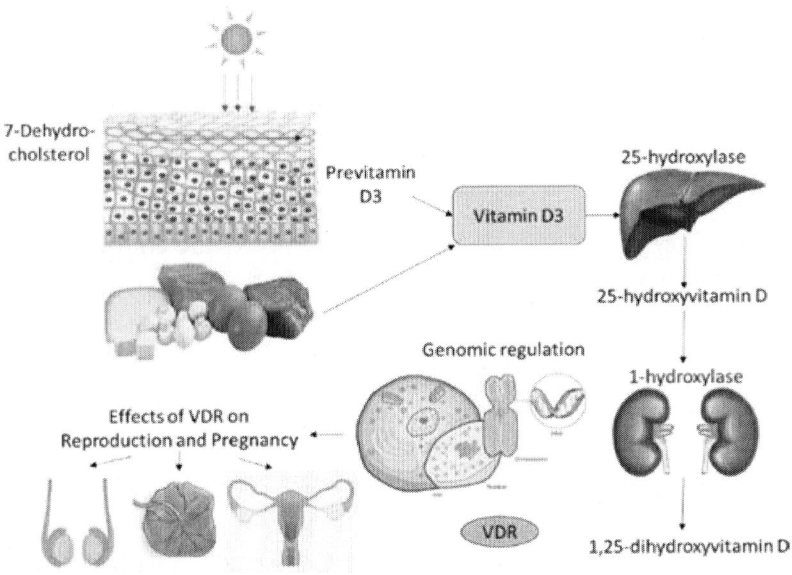

Figure 1. Vitamin D metabolism and effects on reproduction and pregnancy.

VITAMIN D IN REPRODUCTION AND PREGNANCY

Pregnancy leads to notably changes in physical proportions, physiology, requirement and metabolism of vitamin D. Indeed, during these delicate phase of physiological change, the primary role of vitamin D during pregnancy appears to be immunomodulatory, rather than calcium-regulating, although it would also retain that function. Furthermore, vitamin D inadequacy in early life results associated with asthma, multiple sclerosis and other neurological disorders in adulthood, according to the "Barker

Hypothesis" (Heaney 2016) that states that certain adult-onset diseases have their roots in nutritional insults sustained in the perinatal period (either in uterus or in the early months of infancy or both) (Hollis and Wagner 2017).

VITAMIN D A PROGESTERONE-LIKE HORMONE

Vitamin D and its metabolites, from the structural point of view, are called secosteroids, compounds whose structure shows the steroid skeleton with seventeen-carbon tetracyclic hydrocarbon, that is not joined at the level of the bond located in position 9-10 of the ring B. Chemically they are like the mineralocorticoids, glucocorticoids, testosterone, progesterone, estrogens, chorionic gonadotropin (Monastra et al. 2018). The structural similarities between vitamin D and sexual hormones are confirmed by their actions, in particular progesterone and vitamin D play synergistic pivotal functions in female reproductive system.

More in details, they simultaneously act in different phases of the menstrual cycle. During the follicular phase, also called proliferative phase, the follicles inside the ovaries start their development to prepare ovulation. This process occurs from the 1st to the 14th day of the menstrual cycle. The next luteal phase begins after ovulation and lasts normally from the 15th to the 28th day. In this phase, the corpus luteum acts as a transient endocrine gland, which develops in this specific period, and secretes progesterone stimulating a prompt endometrial response. Then, the corpus luteum undergoes to neovascularization. Such process is principally under the control of vascular endothelial growth factor (VEGF) and fibroblast growth factor (FGF), both upregulated in the luteinized granulosa cells, and then absolutely necessary for the embryo implantation (Chakraborty, Das, and Dey 1995; Sak et al. 2013). The corpus luteum produces several different hormones; among them, progesterone is the most important since its activity is necessary and sufficient to hinder the proliferation of endometrium and promote its receptive state for blastocyst implantation to maintain early pregnancy. Therefore, progesterone prepares the environment conducive to pregnancy success and it is categorically necessary, during the secretory phase, to induce the endometrium receptivity to the embryo implantation, even if moderate concentrations are sufficient (Young 2013). In this scenario, vitamin D plays the role of second main actor, it modulates the innate and adaptive immune system, controls cell proliferation and functions, as a regulator of various and important metabolic processes, with effects on the implant, stimulates the production of cytokines and the immune response to infections (Monastra et al. 2018).

VITAMIN D AND PROGESTERONE LEVELS DURING PREGNANCY

Progesterone production starts right after the ovulation and, if pregnancy occurs, it progressively increases during gestation. It is secreted by the corpus luteum until the 12th week, after that, the placenta becomes the principal site of progesterone production (Figure 2).

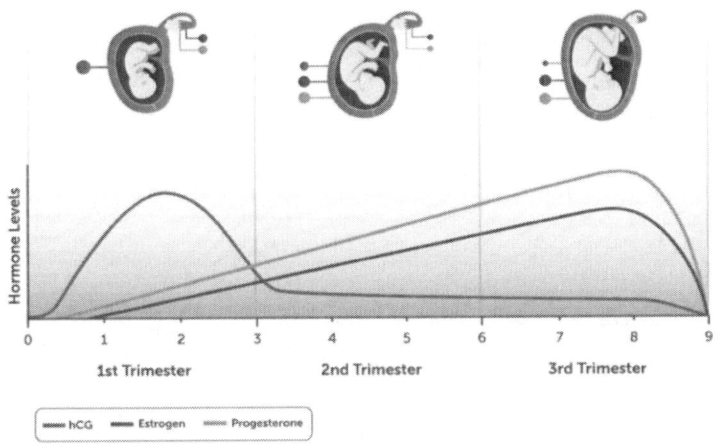

Figure 2. Human chorionic gonadotropin (hCG), progesterone and estrogen production during pregnancy. The production of the three hormones (hCG, progesterone and estrogen) from placenta and ovaries during pregnancy is illustrated. The higher production of each one is rappresented with bold lines and lower production with thinner lines.

Circulating calcidiol (25(OH)D) levels are increased during pregnancy. The placenta is recognized to be the major extra-renal site converting from 25(OH)D to 1,25-dihydroxycholecalciferol (1,25$(OH)_2$D). In addition, the number of VDR increases continuously during the first and second trimester of pregnancy, meaning that this vitamin plays a key physiological role in pregnancy. Especially in the first months of gestation adequate levels of vitamin D are essential, affecting not only the well-being of the mother but also that one of the fetus (Monastra et al. 2018). The presence of VDR has been highlighted both at the level of the decidua and at the level of the trophoblast, suggesting that at this level, the production of 25(OH)D occurs through an autocrine and/or paracrine mechanism.

Immunomodulatory Function in Pregnancy

From the initial implantation of the conceptus, the maternal uterine endometrium undergoes to a 'decidualization' process to support placental development and function. The resulting decidua is a tissue formed from the maternal endometrium, originating from epithelial and stromal cells, and is characterised by invasion from the extraembryonic fetal-derived trophoblasts and close 'cell–cell juxtaposition' of these different tissues. The principal function of the decidua is to facilitate early fetal–maternal exchange of nutrients, gases and waste, while also acting as a secretory source of an array of steroid hormones, cytokines and growth factors (Gellersen, Brosens, and Brosens 2007; Salamonsen, Hannan, and Dimitriadis 2007). However, the decidua also plays a key role in protecting pregnancy against maternal immune surveillance (Warning, McCracken, and Morris 2011). Pregnancy presents a unique immune challenge for the maternal host. Systemically, significant gestation dependent immune adaptions arise throughout pregnancy and, characterization of these, is a major focus of materno-fetal immunology research. Specifically, in the mother, there is a shift towards immune tolerance to accommodate the conceptus (Alijotas-Reig, Llurba, and Gris 2014). As a result, the maternal–fetal interface is a prime area of immune regulation, but cellular immunity within the decidua, is importantly different from the one within the maternal circulation. Cellular infiltration is a key feature of immune function within decidua, and leukocytes comprise at least 40% of the total decidual stromal cell population. The leukocyte subtypes present include decidual (uterine) natural killer (uNK) cells, macrophage subtypes, CD4C and CD8C T-lymphocytes (including T-regulatory cells (Tregs)) and antigen-presenting cells (APCs) such as dendritic cells (DCs) (Arck and Hecher 2013). Although well classified anatomically, these immune cells and, in particular, their functions are the focus of much contemporary research (Oreshkova, Dimitrov, and Mourdjeva 2012).

Immunomodulatory Action of Vitamin D

Pregnancy represents a unique immune challenge for the maternal host, with the maternal–fetal interface being the first area of immune regulation. In this setting, the immune cells within the heterogeneous population of cells that make up the maternal decidua, play a key role in maintaining tolerance of the developing fetus while protecting the conceptus against infection and inflammation. To successfully maintain the pregnancy, the decidual immune system requires a set of regulatory factor and, in this setting, vitamin D may play a highly versatile role by promoting antibacterial innate immune responses to infection, while suppressing adverse inflammatory adaptive immunity.

VDR are found on nearly all immune cells, including lymphocytes (Th1, Th2, Tregs), B-lymphocytes, dendritic cells, Th17 cells, monocytes, and macrophages. The $1,25(OH)_2D_3$, the hormonally active form of vitamin D, modulates the expression, either inhibits or up-regulates, of cytokines in dendritic cells, Th1 cells, Th2 cells, CD8+ cells, T-regulatory cells, B-cells, monocytes, and macrophages.

The $1,25(OH)_2D_3$ decreases the activity of Th1 cells by decreasing the production of interferon gamma (IFN-γ), interleukins (IL-2, IL-12) and tumor necrosis factor (TNF). This effect comes with a complementary increase in the activity of Th2 cells, particularly because of an increase in

IL-10, and a decrease in IL-2 secretion. $1,25(OH)_2D_3$ has also been found to have an inhibitory effect on Th-17 cells. This effect is most likely tied to its inhibitory effect on IL-6 and IL-23 production. Similarly, vitamin D contributes to the activation of T regulatory cells. In the setting of pregnancy, increased T-regulatory cell activity, is essential for the development of a maternal environment in which there is no immune response against the fetus. Vitamin D mediates its stimulatory effect on T-regulatory cells by upregulating forkhead box protein 3 (FOXP3), a factor that supports the differentiation and expansion of T-regulatory cells (Mora, Iwata, and Von Andrian 2008; Shin et al. 2010).

Therefore, supporting the role of vitamin D as a regulator of decidual immunity. Crucially, the placenta is one of the principal sites for extra-renal synthesis of $1,25(OH)_2D$, with both the maternal and fetal sides of the placenta cooperating to maintain high localized tissue levels of this hormone. This is likely to be a key factor in optimizing the immune function of vitamin D, but may also be important in promoting the crosstalk between the maternal decidua and fetal trophoblast. However, it is important to recognize that the success of these decidual responses to vitamin D will be dependent on the availability of substrate 25(OH)D, so that tissue-specific levels of $1,25(OH)_2D$ may be compromised under conditions of vitamin D-insufficiency (Smith et al. 2015). The impact of this on immune function during pregnancy, particularly within the decidua, is likely to be a key component of future studies of vitamin D and pregnancy.

Vitamin D and Progesterone in the Inception of Pregnancy

It was observed that, in response to IL-1β secreted by the blastocyst, the dendritic cells of the endometrium and the macrophages produce 1α-OHase (the enzyme responsible for the last step of the synthesis process) and, consequently, calcitriol (Viganò et al. 2006). On the other hand, the stimulation of the endometrial 1α-OHase expression, due to TNF, is not as potent as that one achieved using very low concentrations of IL-1 β (Viganò et al. 2006). This evidence means that, between the two main pro-inflammatory cytokines, at least in this case, the central role is played by IL-1 β, whereas TNF exerts a secondary function, despite its strong effects in many contexts. Thus, there is not only an increased calcitriol production in the first stages of pregnancy, but also a wide array of new synthetized receptors in the endometrium, ready for binding.

Using blood cells of women with a history of recurrent pregnancy losses (RPL), vitamin D was reported to hinder *in vitro* the proliferation of Th1 cells, and to decrease the production of their cytokines, such as IFN-γ, IL-2 and TNF. On the other hand, vitamin D stimulated the secretion of Th2 cytokines; among them, IL-10 is the most important anti-inflammatory cytokine released under the effect of calcitriol (Ota et al. 2015). Furthermore, *in vitro* vitamin D significantly decreased the cytotoxicity of NK cells (Ota et al. 2015) and it was able to suppress, in the same cells, the secretion of pro-inflammatory cytokines, such as IFN-γ and TNF, and to increase IL-10 release, as well as CSF2, IL-1, IL-6 too, thus modulating the inflammation that normally develops in the initial phase of pregnancy (Evans et al. 2006). Therefore, vitamin D significantly lowers the frequency of Th17 cells in addition to reducing the Th17/Treg ratio in peripheral blood of patients with recurrent miscarriage compared with the control group. The same modulatory effect of inflammation was seen in the induction and upregulation of the suppressive activity of T regulatory (Treg) cells in skin-draining lymph nodes (Gorman et al. 2016). Vitamin D, by binding to VDR on the endometrium, upregulates the target genes, such as calbindin and osteopontin, also called secreted phosphoprotein one (SPP1) (Viganò et al. 2006). Development of receptivity requires changes in endometrial gene expression; endometrial expression of calbindin mRNA was shown to be highly regulated during implantation. Indeed, calbindin participates to the regulation of endometrial receptivity, acting as calcium binding protein. It is well known that the function of reproductive organs and its musculature, is depending on the influx of extracellular calcium or the release of intracellular stores (Luu, Nie, and Salamonsen 2004). Osteopontin is an adhesion protein involved in implantation and decidualization (Johnson et al. 2003). Its action has a direct influence as signal transduction at the uterine-placental interface throughout pregnancy, including the regulation of immune cells behavior and cytokine production. Both, calbindin and osteopontin, are essential for the embryo implantation and for the development of the placenta. Their effects on the endometrium favor the success of pregnancy. Therefore, 1,25$(OH)_2$D is reputed to exert an autocrine/paracrine action aimed at regulating both acquired and innate immune responses at the fetal-maternal interface, as confirmed by the presence of the VDR in the endometrium (Monastra et al. 2018).

Recurrent Pregnancy Loss

RPL is a syndrome caused by multiple etiologies such as anatomical, endocrine, genetic, infectious, immunological, thrombotic and unexplained etiologies hence, an investigation of underlying etiologies is often complicated. When the conventional investigation scheme is applied, up to 60% of women with RPL remain unexplained (Jaslow, Carney, and Kutteh 2010). This suggests a limitation of the current guideline and investigation scheme. Recent studies have indicated that immune inflammatory and thrombotic conditions are two major underlying pathologies for RPL (Kwak-Kim et al. 2010). Approximately 20% of women with RPL have autoimmune conditions, such as antiphospholipid antibody (APA) (Ford and Schust 2009), and other autoimmune abnormalities, for instance antinuclear antibody (ANA), anti-thyroperoxidase antibody and anti-thyroglobulin antibody (Thangaratinam et al. 2011; Ticconi et al. 2010). These autoantibodies may be associated with immune inflammatory changes at maternal fetal junction and contribute to pregnancy losses. Moreover, the presence of autoimmunity such as APA or anti-prothrombin antibody may contribute thrombotic changes at the maternal-fetal junction, in addition to inherited/acquired thrombophilia in women with RPL. Therefore, autoimmunity is keenly associated with immune inflammatory and thrombotic pathology of RPL. Recently, vitamin D has been reported to play a major role in auto- and cellular immune responses (Kwak-Kim and et al. 2016).

Vitamin D and Autoimmunity in Women with RPL

Autoimmunity, such as APA, ANA and anti-thyroid antibodies, is often associated with reproductive failures such as RPL or RIF. A significant proportion of women with RPL and autoimmunity have cellular immune abnormalities, including increased NK cell levels, NK cytotoxicities and Th1/Th2 cell ratios. Therefore, the presence of autoimmunity may directly cause reproductive failures or indirectly function as markers or intermediaries for an underlying inflammatory cellular immunity involved in RPL or RIF. As stated before, vitamin D has immune regulatory effects on T and B cell immunity and NK cell cytotoxicity. Vitamin D deficiency is prevalent in women with RPL and women with Vitamin D deficiency have increased auto- and cellular immune abnormalities as compared with women with normal vitamin D levels. In women with low vitamin D, the prevalence of auto-antibodies is higher which is supported by increased CD19+ B cell population and dysregulated B cell subsets. Vitamin D suppresses NK cytotoxicity through reduction of conjugation with target cells, and pro-inflammatory cytokine secretion (Ota et al. 2015). Vitamin D effect on type 2 shift in NK cells is via TLR4/NF-kB pathway. Hence, it is possible that appropriate levels of vitamin D and its supplementation, may have a pivotal role to maintain and succeed in early pregnancy. Potential therapeutic application of vitamin D for RPL with auto- and cellular immune abnormalities should be explored (Kwak-Kim and et al. 2016).

Studies on Vitamin D and Pregnancy Loss

Recent studies demonstrate a link between vitamin D deficiency and pregnancy loss which is possibly mediated by effector CD4+ T helper cellular responses in the innate (Th1) and adaptive (Th2) immune systems. In a retrospective cross sectional study of 133 women diagnosed with RPL, Ota et al. reported that 63 out of 133 women (47.4%) had hypovitaminosis D (less than 30 ng/ml) and that NK cell levels and NK cytotoxicity at effector to target cell (E:T) ratio of 25:1 were significantly higher in the low vitamin

D group when compared with subjects with normal vitamin D levels (Ota et al. 2014). Additionally, in a retrospective cohort study of 115 women with antiphospholipid syndrome (APS), the median levels of vitamin D were significantly lower compared to healthy volunteers (n=128), although the authors report possible confounding factors related to provider recommendation of increased sunscreen use in the APS cohort (Andreoli et al. 2012). Tavakoli et al. reported similar cytokine profiles and vitamin D expression of whole endometrial cells obtained from women with unexplained RPL (n=8) compared to age matched controls (n=8), although the authors do not address the limitations of the small sample size (Tavakoli et al. 2011). A recent study evaluated the effect of vitamin D treatment with and without low molecular weight heparin on an antiphospholipid antibody-induced inflammatory response of human first trimester trophoblast (HTR8) and primary trophoblast cell lines (Gysler et al. 2015). This study reported a decrease in the inflammatory response with vitamin D treatment by measurement of IL-8. Also of note, vitamin D has been associated with increased risk of preeclampsia, infants that are small for gestational age (SGA), cesarean section, and gestational diabetes (Pacis et al. 2014).

Preterm Birth

Preterm birth (PTB) is the birth of a baby before 37 complete weeks of gestation. Every year, 15 million neonates worldwide are born preterm. PTB, as the largest cause of neonatal deaths worldwide (Blencowe et al. 2012), puts surviving children at risk for cerebral palsy, behavioral problems, bronchopulmonary dysplasia, retinopathy of prematurity, hearing impairments, increased hospital admissions and academic underachievement (Saigal and Doyle 2008). The implications of being born too soon extend beyond the neonatal period throughout the life cycle. There are two-way linkages between PTB, low birthweight and non-communicable diseases (NCDs), such as diabetes and hypertension. Firstly, women with these NCDs have an elevated risk of having a low birthweight baby due to prematurity or other causes, demanding increased attention to maternal health and care, including the antenatal diagnosis and management of NCDs (Howson et al. 2013). Premature babies, in turn, are at greater risk of developing NCDs like hypertension and diabetes later in life and, if female, of having a preterm and/or low birthweight baby herself. Thus, prematurity not only affects a newborn directly but can also result in a vicious intergenerational cycle of risk (Howson et al. 2013). The link between prematurity and an increased risk of hypertension, diabetes and other NCDs takes on an added public health importance when the reported increases in the rates of NCDs worldwide are taken into consideration. Currently, nine million people under the age of 60 years die from NCDs per year, accounting for more than 63% of all deaths, with the greatest burden in Africa and other low-income regions, where PTB rates are also high (United Nations General Assembly 2011). With pregestational diabetes and hypertension reported to increase the risk of having a preterm delivery in the US by 38% and 33%, respectively, it is clear that the problem of PTB should be a major concern to policymakers, donor organizations and other stake holders in the NCD as well as reproductive, maternal, newborn and child health (RMNCH) communities (Sibai et al. 2000).

Romero has defined PTB as a syndrome in which different physio-pathological circuits come to play, first of all (according to spontaneous PTBs) infectious inflammatory circuits plays an important role, with the activation of the inflammatory cascade and finally the production of prostaglandins and metalloproteinase that determine: remodeling at the cervical level, rupture of the membranes and the appearance of contractions (Romero, Dey, and Fisher 2014) (Figure 3).

Infection not only initiates inflammatory responses but also was reported to be responsible for 40% of the premature births (Thota, Laknaur, and Farmer 2014).

Lipopolysaccharide (LPS), a bacterial endotoxin has been reported to increase the cytokine secretion in human myometrial cells, enhance prostaglandin production in rodent and human myometrium (Thota et al. 2013) and cause preterm delivery in rodents (Elovitz et al. 2003). In turn, cytokine treatments have been reported to increase prostaglandins, connexin, oxytocin, and oxytocin receptor, which suggests that infection increases cytokines that, in turn, increase contractile-associated proteins and initiate pre-term delivery (Thota, Laknaur, and Farmer 2014).

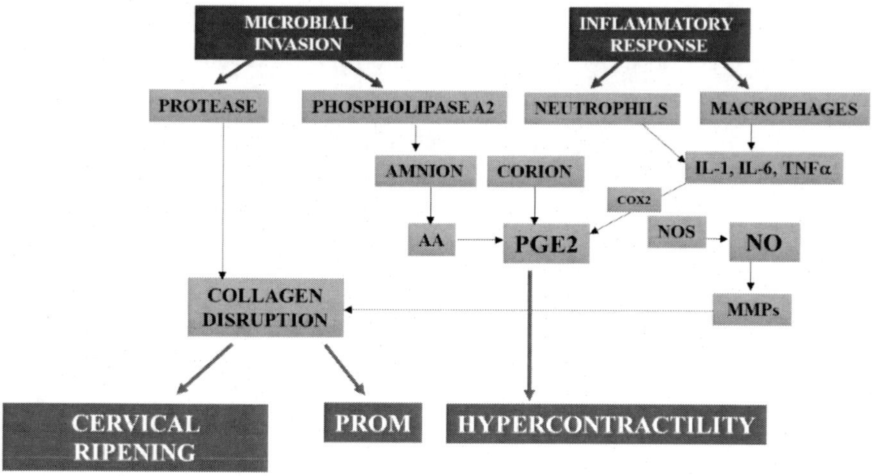

Figure 3. The Role of Inflammation in Preterm birth.

Randomized controlled trials on Vitamin D Supplementation and PTB

In order to give an evidence based indication on the importance of vitamin D supplementation during pregnancy to reduce the risk of PTB, we present the results of several randomized controlled trial (RCTs), observational studies and recent meta-analyses that evaluated whether low maternal circulating 25(OH)D is associated with an increased risk of PTB or spontaneous PTB (sPTB) and whether vitamin D supplementation alone during pregnancy, can reduce the risk of PTB.

A recent meta-analysis by Zhou et al, included twenty-four articles (six RCTs and 18 observational studies). The authors reported that maternal circulating 25(OH)D deficiency, rather than insufficiency, was significantly associated with an increased risk of PTB, and vitamin D supplementation alone during pregnancy could reduce the risk of PTB (pooled RR, 0.57; 95%

CI: 0.36–0.91). This was also the case for the sPTB subgroup, in which circulating 25(OH)D < 50 compared with >50 nmol/L resulted in a pooled OR, 1.45; 95%CI: 1.20–1.75. Thus, they concluded that maternal circulating 25(OH)D deficiency could increase PTB risk, and vitamin D supplementation alone during pregnancy could reduce PTB risk (Zhou et al. 2017).

These results are confirmed by another meta-analysis of Qin et al. that included 10 studies with a total of 10,098 participants. They found that pregnant women with vitamin D deficiency experienced a significantly increased risk of PTB (OR = 1.29; 95% CI:1.16-1.45) with low heterogeneity (I2 = 25%; p = 0.21). Sensitivity analysis showed that, exclusion of any single study, did not materially alter the overall combined effect. The authors concluded that pregnant women with vitamin D deficiency during pregnancy have an increased risk of PTB (Qin et al. 2016).

A study examined vitamin D action on myometrial cells simulating an *in vivo* inflammatory environment by co-culturing uterine smooth muscle cells with monocyte lineage (THP1) cells and assess whether vitamin D regulates pro-inflammatory cytokines, contractile-associated proteins, steroid hormone receptor expression, and NFkB pathway proteins in human uterine myometrial smooth muscle (UtSM) cells. The author found that vitamin D treatment significantly decreased the expression of inflammatory markers and contractile-associated factors in UtSM cells. Similarly, vitamin D decreased estrogen receptor α and progesterone receptors A to B ratio in UtSM cells, that were co-cultured with THP1 cells. In addition, vitamin D treatment significantly decreased monocyte induced p-IkBa in cytosol and NFkB-p65 in the nucleus and increased IkBa in cytosol in UtSM cells. These results suggest that vitamin D treatment decreases inflammation-induced cytokines and contractile-associated factors in the uterine myometrial smooth muscle cells through the NFkB pathway (Thota, Laknaur, and Farmer 2014).

Table 1. Summary of findings for the main comparison (De-Regil et al. 2016)

Outcomes	Anticipated absolute effects (95% CI)		Relative effect (95% CI)	N	Quality of the evidence (GRADE)
	Risk with no treatment/placebo	*Risk with vitamin D alone*			
Pre-eclampsia	*Study population*		RR 0.52 (0.25-1.05)	219 (2 RTCs)	LOW
	155 per 1000	80 per 1000			
Gestational diabetes	*Study population*		RR 0.43 (0.05-3.45)	219 (2 RTCs)	VERY LOW
	24 per 1000	10 per 1000			
Preterm birth (<37 weeks)	*Study population*		RR 0.36 (0.14-0.93)	477 (3 RTCs)	MODERATE
	99 per 1000	36 per 1000			
Low birth weight (<2500 g)	*Study population*		RR 0.40 (0.24-0.67)	493 (3 RTCs)	MODERATE
	199 per 1000	80 per 1000			

Last but not least, an overview of Cochrane systematic reviews analyzed the efficacy of the interventions during pregnancy to prevent PTB (Medley et al. 2018). They included a systematic review by De-Regil et al., which evaluated vitamin D supplementation for women during pregnancy (De-Regil et al. 2016). The authors included 15 trials assessing a total of 2833 women. Nine trials compared the effects of vitamin D alone versus no supplementation or a placebo and six trials compared the effects of vitamin D and calcium with no supplementation. Vitamin D alone, versus no supplementation or placebo, consistently show that women who received vitamin D supplements alone, particularly on a daily basis, had higher 25(OH)D than those receiving no intervention or placebo. Data from three trials involving 477 women suggest that vitamin D supplementation during pregnancy reduces the risk PTB compared to no intervention or placebo (8.9% versus 15.5%; RR 0.36; 95% CI 0.14 to 0.93, moderate quality) (Table 1). Surprisingly, women who received vitamin D with calcium had an increased risk of PTB (RR 1.57; 95% CI 1.02 to 2.43, three studies, 798 women, moderate quality). They concluded that supplementing pregnant women with vitamin D in a single or continued dose increases serum 25(OH)D at term and may reduce the risk of preeclampsia, low birthweight and PTB. However, when vitamin D and calcium are combined, the risk of

PTB is increased. The clinical significance of the increased serum 25(OH)D concentrations is still unclear. In light of this, these results need to be interpreted with caution. The evidence on whether vitamin D supplementation should be given as a part of routine antenatal care to all women to improve maternal and infant outcomes remains unclear. While there is some indication that vitamin D supplementation could reduce the risk of preeclampsia and increase length and head circumference at birth, further rigorous RCTs are required to confirm these effects (De-Regil et al. 2016).

CONCLUSION

Pregnancy presents a concrete immune challenge for the maternal host, with the maternal–fetal interface being a critical area of immune regulation. The concert of actions that must occur, in order to avoid maternal rejection of the foreign tissue of the fetus, is substantial. There must be a perfect balance between maternal tolerance to these foreign proteins of paternal origin, but also immune surveillance and function such that the mother is not immunocompromised. Indeed, the immune cells within the heterogeneous population of cells make up the maternal decidua and play a key role in maintaining tolerance of the developing fetus while protecting the conceptus against infection and inflammation. Successful maintenance of this complex decidual immune system requires an equally complex set of regulatory factor and, in this setting, vitamin D may play a highly versatile role by promoting antibacterial innate immune responses to infection, while suppressing adverse inflammatory adaptive immunity. When this process goes awry, the mother may experience miscarriage or pregnancy complications as preeclampsia and infections. Vitamin D deficiency affects these processes. The high prevalence of vitamin D deficiency during pregnancy and its association with PTB have attracted much public health attention. PTB is a heterogeneous syndrome, most commonly caused by infection and inflammation. Vitamin D, acting as an immune modulator, can reduce the possibility of PTB by inhibiting inflammation and regulating immune function.

There is a growing need for future research endeavors to focus not only on critical period(s), from pre-conception through pregnancy, but throughout life. There is urgency based on emerging research to correct vitamin D deficiency and maintain optimal vitamin D status. The impact of vitamin D and its metabolites on genetic signaling during pregnancy in both mother and fetus is an area of great activity and still in its early stages. While vitamin D repletion during pregnancy minimizes the risk of certain adverse outcomes (e.g., PTB, asthma, preeclampsia, and gestational diabetes), the mechanisms of how these processes occur are not fully understood.

The impact of this on immune function during pregnancy, particularly within the decidua, is likely to be a key component of future studies of vitamin D and pregnancy.

References

Alijotas-Reig, J, E Llurba, and JM Gris. 2014. "Potentiating Maternal Immune Tolerance in Pregnancy: A New Challenging Role for Regulatory T Cells." *Placenta* 35: 241–248.

Andreoli, L, S Piantoni, F Dall'Ara, F Allegri, PL Meroni, and A Tincani. 2012. "Vitamin D and Antiphospholipid Syndrome." *Lupus* 21 (7): 736–40.

Arck, PC, and K Hecher. 2013. "Fetomaternal Immune Cross-Talk and Its Consequences for Maternal and Offspring's Health." *Nature Medicine* 19: 548–556.

Blencowe, H, S Cousens, MZ Oestergaard, D Chou, AB Moller, R Narwal, A Adler, C Vera Garcia, S Rohde, and L Say. 2012. "National, Regional, and Worldwide Estimates of Preterm Birth Rates in the Year 2010 with Time Trends since 1990 for Selected Countries: A Systematic Analysis and Implications." *Lancet* 379: 2162–2172.

Chakraborty, I, SK Das, and SK Dey. 1995. "Differential Expression of Vascular Endothelial Growth Factor and Its Receptor MRNAs in the Mouse Uterus around the Time of Implantation." *J Endocrinol* 147: 339–52.

Chen, YH, L Fu, JH Hao, H Wang, C Zhang, FB Tao, and DX Xu. 2018. "Influent Factors of Gestational Vitamin D Deficiency and Its Relation to an Increased Risk of Preterm Delivery in Chinese Population." *Scientific Reports* 8 (3608).

De-Regil, LM, C Palacios, LK Lombardo, and JP Peña-Rosas. 2016. "Vitamin D Supplementation for Women during Pregnancy." *Cochrane Database Syst Rev* 14 (1).

Elovitz, MA, Z Wang, EK Chien, DF Rychlik, and MA Phillippe. 2003. "New Model for Inflammationinduced Preterm Birth: The Role of Plateletactivating Factor and Toll-like Receptor-4." *Am J Pathol* 163: 2103–11.

Evans, KN, L Nguyen, J Chan, BA Innes, JN Bulmer, MD Kilby, and M Hewison. 2006. "Effects of 25-Hydroxyvitamin D3 and 1,25-Dihydroxyvitamin D3 on Cytokine Production by Human Decidual Cells." *Biol Reprod* 75: 816–22.

Ford, HB, and DJ Schust. 2009. "Recurrent Pregnancy Loss: Etiology, Diagnosis, and Therapy." *Rev Obstet Gynecol* 2: 76–83.

Gellersen, B, IA Brosens, and JJ Brosens. 2007. "Decidualization of the Human Endometrium: Mechanisms, Functions, and Clinical Perspectives." *Seminars in Reproductive Medicine* 25: 445–453.

Gorman, S, S Geldenhuys, M Judge, CE Weeden, J Waithman, and PH Hart. 2016. "Dietary Vitamin D Increases In, Percentages and Function of Regulatory T Cells Suppresses, the Skin-Draining Lymph Nodes and 2016;, Dermal Inflammation." *J Immunol Res*, 1426503.

Gysler, SM, MJ Mulla, M Stuhlman, AK Sfakianaki, MJ Paidas, and NL Stanwood. 2015. "Vitamin D Reverses APL-Induced Inflammation and LMWH-Induced SFlt-1 Release by Human Trophoblast." *Am J Reprod Immunol* 73 (3): 242–50.

Halloran, BP, and HF Deluca. 1980. "Effect of Vitamin-D Deficiency on Fertility and Reproductive Capacity in the Female Rat." *J Nutr* 110: 1573–80.

Heaney, RP. 2016. "Is Vitamin D Inadequacy in Early Life an Instance of the 'Barker Hypothesis'?" *Nutr Today* 51: 14–17.

Hollis, BW, CL Wagner. 2017. "New Insights into the Vitamin D Requirements during Pregnancy." *Bone Research* 5: 17030.

Howson, CP, MV Kinney, L McDougall, and JE Lawn. 2013. "Born Too Soon: Preterm Birth Matters." *Reproductive Health* 10 (1): S1.

Irani, M, Z Merhi. 2014. "Role of Vitamin D in Ovarian Physiology and Its Implication in Reproduction: A Systematic Review." *Fertil Steril* 102 (2): 460-468.e3.

Jaslow, CR, JL Carney, and WH Kutteh. 2010. "Diagnostic Factors Identified in 1020 Women with Two versus Three or More Recurrent Pregnancy Losses." *Fertil Steril* 290 (93): 1234–1243.

Johnson, GA, RC Burghardt, FW Bazer, and TE Spencer. 2003. "Osteopontin: Roles in Implantation and Placentation." *Biol Reprod* 69: 1458–71.

Kwak-Kim, J, and et al. 2016. "Humoral and Cellular Autoimmunity in Women with Recurrent Pregnancy Losses and Repeated Implantation Failures: A Possible Role of Vitamin D." *Autoimmun Rev*.

Kwak-Kim, J, JC Park, HK Ahn, JW Kim, and A Gilman-Sachs. 2010. "Immunological Modes of Pregnancy Loss." *Am J Reprod Immunol* 63: 611–623.

Luu, KC, GY Nie, and LA Salamonsen. 2004. "Endometrial Calbindins Are Critical for Embryo Implantation: Evidence from in Vivo Use of Morpholino Antisense Oligonucleotides." *Proc Natl Acad Sci USA* 101: 8028–33.

Medley, N, JP Vogel, A Care, and Z Alfirevic. 2018. "Interventions during Pregnancy to Prevent Preterm Birth: An Overview of Cochrane Systematic Reviews." *Cochrane Database Syst Rev* 14 (11): CD012505.

Monastra, G, S De Grazia, L De Luca, S Vittorio, and V Unfer. 2018. "Vitamin D: A Steroid Hormone with Progesterone-like Activity." *European Review for Medical and Pharmacological Sciences* 22: 2502–12.

Mora, JR, M Iwata, and UH Von Andrian. 2008. "Vitamin Effects on the Immune System: Vitamins A and D Take Centre Stage." *Nat Rev Immunol* 8: 685–698.

Mumford, SL, RA Garbose, K Kim, K Kissell, DL Kuhr, UR Omosigho, NJ Perkins, et al. 2018. "Association of Preconception Serum 25-Hydroxyvitamin D Concentrations with Livebirth and Pregnancy Loss: A Prospective Cohort Study." *Lancet Diabetes Endocrinol.*

Oreshkova, T, R Dimitrov, and M Mourdjeva. 2012. "A Cross-Talk of Decidual Stromal Cells, Trophoblast, and Immune Cells: A Prerequisite for the Success of Pregnancy." *American Journal of Reproductive Immunology* 68: 366–373.

Ota, K, S Dambaeva, AR Han, K Beaman, A Gilman-Sachs, and Kim Kwak-J. 2014. "Vitamin D Deficiency May Be a Risk Factor for Recurrent Pregnancy Losses by Increasing Cellular Immunity and Auto-immunity." *Hum Reprod* 29 (2): 208–19.

Ota, K, S Dambaeva, MW Kim, AR Han, A Fukui, A Gilman-Sachs, K Beaman, and J Kwak-Kim. 2015. "1,25-Dihydroxy- Vitamin D3 Regulates NK-Cell Cytotoxicity, Cytokine Secretion, and Degranulation in Women with Recurrent Pregnancy Losses." *Eur J Immunol* 45: 3188–99.

Ozkan, S, S Jindal, K Greenseid, J Shu, G Zeitlian, C Hickmon, and Pal L. 2010. "Replete Vitamin D Stores Predict Reproductive Success Following in Vitro Fertilization." *Fertil Steril* 94: 1314–1319.

Pacis, MM, CN Fortin, SM Zarek, JH Mumford, and JH Segars. 2014. "Vitamin D and Assisted Reproduction: Should Vitamin D Be Routinely Screened and Repleted Prior to ART? A Systematic Review." *J Assist Reprod Genet*. https://doi.org/DOI: 10.1007/s10815-014-0407-9.

Qin, L, F Lu, S Yang, H Xu, and B Luo. 2016. "Does Maternal Vitamin D Deficiency Increase the Risk of Preterm Birth. A Meta -Analysis of Observational Studies." *Nutrients* 8 (5): 301.

Romero, R, SK Dey, and SJ Fisher. 2014. "Preterm Labor: One Syndrome, Many Causes." *Science* 15345: 760–5.

Saigal, S, and LW Doyle. 2008. "An Overview of Mortality and Sequelae of Preterm Birth from Infancy to Adulthood." *Lancet* 371: 261–269.

Sak, ME, T Gul, MS Evsen, HE Soydinc, S Sak, A Ozler, and U Alabalik. 2013. "Fibroblast Growth Factor-1 Expression in the Endometrium of

Patients with Repeated Implantation Failure after in Vitro Fertilization." *Eur Rev Med Pharmacol Sci* 17: 398–402.

Salamonsen, LA, NJ Hannan, and E Dimitriadis. 2007. "Cytokines and Chemokines during Human Embryo Implantation: Roles in Implantation and Early Placentation." *Seminars in Reproductive Medicine* 25: 437–444.

Shin, J, M Choi, M Longtine, and D Nelson. 2010. "Vitamin D Effects on Pregnancy and the Placenta." *Placenta* 31: 1027–1034.

Sibai, BM, SN Caritis, JC Hauth, C MacPherson, JP VanDorsten, M Klebanoff, and et al. Landon M, Paul RH, Meis PJ, Miodovnik M. 2000. "Preterm Delivery in Women with Pregestational Diabetes Mellitus or Chronic Hypertension Relative to Women with Uncomplicated Pregnancies. The National Institute of Child Health and Human Development Maternal-Fetal Medicine Units Network." *Am J Obstet Gynecol* 183: 1520-1524.

Smith, TA, DR Kirkpatrick, O Kovilam, and DK Agrawal. 2015. "Immunomodulatory Role of Vitamin D in the Pathogenesis of Preeclampsia." *Expert Rev Clin Immunol.* 11 (9): 1055–1063.

Sollis, Susanne Stoddard. 2015. "Vitamin D Deficiency and Infertility: A Systematic Review." *All Student Publications* 143.

Tamblyn, JA, M Hewison, CL Wagner, JN Bulmer, and MD Kilby. 2015. "Immunological Role of Vitamin D at the Maternal–Fetal Interface." *Journal of Endocrinology* 224: R107–R121.

Tavakoli, M, M Jeddi-Tehrani, A Salek-Moghaddam, S Rajaei, A Mohammadzadeh, and S Sheikhhasani. 2011. "Effects of 1,25(OH)2 Vitamin D3 on Cytokine Production by Endometrial Cells of Women with Recurrent Spontaneous Abortion." *Fertil Steril* 96 (3): 751–7.

Thangaratinam, S, A Tan, E Knox, MD Kilby, J Franklyn, and A Coomarasamy. 2011. "Association between Thyroid Autoantibodies and Miscarriage and Preterm Birth: Meta-Analysis of Evidence." *BMJ* 342: d2616.

Thota, C, T Farmer, RE Garfield, R Menon, and A Al-Hendy. 2013. "Vitamin D Elicits Anti-Inflammatory Response, Inhibits Contractile-

Associated Proteins, and Modulates Toll-like Receptors in Human Myometrial Cells." *Reprod Sci* 20: 463–75.

Thota, C, A Laknaur, and T Farmer. 2014. "Vitamin D Regulates Contractile Profile in Human Uterine Myometrial Cells via NF-KB Pathway." *Am J Obstet Gynecol* 210 (347): e1-10.

Ticconi, C, F Rotondi, M Veglia, A Pietropolli, S Bernardini, and F Ria. 2010. "Antinuclear Autoantibodies in Women with Recurrent Pregnancy Loss." *Am J Reprod Immunol*, no. 64: 384–392.

United Nations General Assembly. 2011. "*Political Declaration of the High-Level Meeting of the General Assembly on the Prevention and Control of Noncommunicable Disease. Sixty-Sixth Session.*"

Viganò, P, D Lattuada, S Mangioni, L Ermellino, M Vignali, E Caporizzo, P Panina-Bordignon, M Besozzi, and AM Di Blasio. 2006. "Cycling and Early Pregnant Endometrium as a Site of Regulated Expression of the Vitamin D System." *Journal of Molecular Endocrinology* 36 (3): 415–424.

Wacker, M, and MF Holick. 2013. "Vitamin D Effects on Skeletal and Extra Skeletal Health and the Need for Supplementation." *Nutrients* 5 (1): 111–48.

Wagner, CL, C Baggerly, S McDonnell, KA Baggerly, CB French, L Baggerly, SA Hamilton, and BW Hollis. 2016. "Post-Hoc Analysis of Vitamin D Status and Reduced Risk of Preterm Birth in Two Vitamin D Pregnancy Cohorts Compared with South Carolina March of Dimes 2009-2011 Rates." *J Steroid Biochem Mol Biol* 155: 245–51.

Warning, JC, SA McCracken, and JM Morris. 2011. "A Balancing Act: Mechanisms by Which the Fetus Avoids Rejection by the Maternal Immune System." *Reproduction* 141: 715–724.

Young, SL. 2013. "Oestrogen and Progesterone Action on Endometrium: A Translational Approach to Understanding Endometrial Receptivity." *Reprod Biomed Online* 27: 497–505.

Zhou, SS, YH Tao, K Huang, BB Zhu, and FB Tao. 2017. "Vitamin D and Risk of Pretermbirth: Up-to-Date Meta-Analysis of Randomized Controlled Trials and Observational Studies." *J. Obstet. Gynaecol. Res* 43 (2): 247–256.

In: Vitamin D as Progesterone-Like Hormone ISBN: 978-1-53617-516-5
Editor: Vittorio Unfer

Chapter 5

VITAMIN D WITH ALPHA-LIPOIC ACID AND MAGNESIUM FOR CONTROLLING THE UTERINE CONTRACTILE ACTIVITY

Imelda Hernández Marín*[*]*MD, Liliana Velázquez, MD and Juan Rivero, MD
Biología de la Reproducción Humana,
Universidad Nacional Autónoma de México,
Ciudad de México, México

ABSTRACT

The overall rate of preterm birth is constantly increasing worldwide. In most developed countries, the incidence may range from 5% to 10%. One of the primary symptoms related to preterm birth are premature uterine contractions. Uterine contraction can be treated and/or inhibited with tocolytics drugs. Among them, in obstetrics the most frequently used are oxytocin-receptor antagonists, inhibitors of prostaglandin synthesis,

---

[*] Corresponding Author's Email: marime64@hotmail.com.
Biología de la Reproducción Humana, Universidad Nacional Autónoma de México, Ciudad México, México

calcium antagonists, nitric oxide (NO) donors, betamimetics and magnesium sulfate. These drugs may present important side effects that should be considered when choosing the right agent. Recently it was shown that oral supplementation with magnesium and lipoic acid reduces the incidence of premature uterine contraction and consequent events of hospitalization. Administration of vitamin D also was shown to reduce the risk of pre-eclampsia, low birthweight and preterm birth. Therefore, it would be useful to better investigate on the combination of natural molecules, such as vitamin D, magnesium and lipoic acid in controlling the uterine contractile activity.

Keywords: uterine activity, premature delivery, preterm birth, vitamin D, alpha lipoic acid and magnesium

INTRODUCTION

The World Health Organization (WHO) defines preterm birth when occurring before the 37 completed weeks of gestation or less than 259 days from the first day of the last menstrual period in women.

Preterm birth is considered a concern worldwide. Its incidence is approximately 15 million events per year (Griffin et al., 2019). In most developed countries, the incidence may range from 5% to 10%. It is considered as the main cause of neonatal morbidity and mortality, and it is not only estimated as an emotional burden for families, but also an economic burden for society. The average hospital stay for premature newborns is 13 days compared to 1.5 days for full-term newborns, and the medical costs of premature babies in the first year of life are ten times higher than the full-term newborns. Furthermore, preterm birth is a leading cause of neonatal mortality. Premature babies are at risk of multiple complications, and neonatal outcomes are closely related to gestational age at birth (Griffin et al., 2019). In addition, it is characterized by the presence of uterine contractions associated with progressive cervical shortening, dilation, premature activation of fetal membranes as well as a cervical collagen remodeling (Vogel et al., 2018). The recent reduced incidence of preterm birth in the United States may be due to the newer recommendations from the American College of Obstetricians and Gynecologists that consist on progesterone treatment for women at high risk of preterm birth and stricter guidelines for assisted reproductive technology (ACOG, 2012; ACOG, 2013). Furthermore, elective or nonmedically indicated deliveries before 39 weeks' gestation have been reduced by the implementation of an improvement program (Oshiro et al., 2013).

ETIOLOGY OF PRETERM BIRTH

The etiology of preterm birth is multifactorial. There is a wide range of causal factors such as sociodemographic, medical, obstetric, fetal, and environmental. Approximately two thirds of premature births occur without a clear risk factor. An inflammatory response is always involved in the activation of kappa nuclear factor of activated B-cell enhancer light chain (NF-kB) that determines the synthesis and release of inflammatory molecules, involved in the activation of fetal membranes. In most cases, the triggering cause (idiopathic cause) is unknown; however, multiple gestation and polyhydramnios are identified as risk factors. One of the main causes might be the occurrence of infections such as pyelonephritis or appendicitis; however, the most frequent cause is subclinical intra-amniotic infection and/or inflammation that cause 18% of the total premature deliveries (Purisch et al., 2017). Indeed, inflammation and infection are considered the major predisposing factors for preterm birth. It has been demonstrated that elevated Interleukin 6 (IL-6) in cervicovaginal fluid and C-reactive protein (CRP) in amniotic fluid are correlated with a higher risk of spontaneous preterm birth among asymptomatic women (Wei et al. 2010).

One of the primary symptoms related to preterm birth are premature uterine contractions. In normal conditions, during labor, uterine contractions are the tightening and shortening of the uterine muscles, that cause the cervix to dilate and efface and help the baby to descend into the birth canal. When regular contractions of the uterus, resulting in changes in the cervix, occur before the 37th week of pregnancy, this increases the risk of preterm labor. For the prevention of preterm labor, it is important an early identification of symptoms and the use of pharmacological treatments. Uterine contractions can be treated and/or inhibited with tocolytics drugs, thus, preventing preterm birth. Among them, in obstetrics the most frequently used are calcium antagonists, oxytocin receptor antagonists, inhibitors of the synthesis of prostaglandins, betamimetics and magnesium sulfate (Carson, 2004; Groom, 2007; Hubinont and Debieve, 2011). Most of these drugs have unavoidable and important side effects that must be taken into account when choosing the ideal therapeutic agent (Purisch et al., 2017).

The main neonatal complications of premature delivery include respiratory distress syndrome, sepsis, intraventricular hemorrhage, necrotizing enterocolitis, hypothermia, hypoglycemia, hyperbilirubinemia and difficulty in feeding. Long-term morbidity includes retinopathy of prematurity, neurodevelopmental disorders and cerebral palsy; therefore, preventive and therapeutic measures and treatments are currently of great importance in the face of the obvious global public health problem that is considered.

PATHOPHYSIOLOGY OF UTERINE CONTRACTIBILITY

The uterus is composed of a contractile unit called myometrium, which is made up of two layers of smooth muscle, a circular inner layer and a longitudinal outer layer. These have the ability to contract without nerve or hormonal stimuli. The electrical activity of this unit is controlled by changes in the action potential of the smooth muscle cell membrane (Rabotti and Mischi, 2014).

The regulation of uterine activity depends on several factors, including progesterone, relaxin, prostacyclin (PGI2), the peptide related to parathyroid hormone (PTH-rP), nitric oxide and corticotropin-releasing hormone (CRH) that can act as inhibitory or stimulators factors of uterine activity (Fomin et al., 2006). The activation of uterine activity mainly involves the following pathways:

1. Adrenal pituitary hypothalamus axis.

2. Mechanical distension of the uterus.
3. Oxytocin stimulates uterine contractions through 2 mechanisms of action:
   (a) Stimulation of phospholipase C (PLC), that acts as a second messenger to increase the release of intracellular calcium.
   (b) Opening the calcium channels, hence, increasing the concentration of intracellular calcium (Fomin et al., 2006).

In parvocellular neurons of the paraventricular nucleus there is an increase in the synthesis of propiomelanocortin mRNA (POCM) causing an increase in adrenocorticotropic hormone (ACTH) concentrations in the fetal circulation; this acts on the fetal adrenal gland activating steroidogenesis in order to increase cortisol production. This mechanism occurs mainly during hypoxia.

CRH stimulates the synthesis of prostaglandins, $PGF_{2\alpha}$ and PGE_2, in amniotic, chorionic, deciduous, placental and myometrial cells, therefore increasing the bioavailability of prostaglandins and inflammatory markers in the maternal circulation such as interleukins (IL-6 and IL-8) and tumor necrosis factor α (TNFα) related to the presence of premature delivery (Rood and Buhimschi, 2017).

Finally, the toll 4 receptor (TLR4) is a member of the TLR family (Toll type receptors) that is available in the placenta in the maternal and fetal face, as well as in lymphocytes; its expression increases throughout pregnancy, acting as a first-line rapid response system if a systemic infection occurs.

Lipopolysaccharide (LPS) is a bacterial endotoxin that can activate TLR4; the increase in LPS concentrations leads to chronic inflammation by increasing cytokine secretion and the production of prostaglandins in myometrial cells. Early activation of this pathway can lead to a breakdown of the maternal - fetal immune balance by playing a critical role in inducing preterm labor.

In addition, infections induce phosphorylation of inhibitor of kappa B alpha (IκBα), a protein that inactivates the nuclear factor kappa-light-chain-enhancer of activated B cells (NF-κB), causing translocation of NF-κB, which increase the proinflammatory cytokines, enhancing the expression of contractile uterine factors (Singh et al., 2015).

Multiple studies have been conducted to locate the pacemaker area of the uterus. In 1952, Caldeyro-Barcia and Álvarez described the uterine fundus as the site of onset of uterine activity (Álvarez and Caldeyro-Barcia, 1954). However, in recent experimental studies it has been observed that the uterine horn is one of the main sites of onset of uterine activity, although the concept of multiple pacemakers has been proposed by different authors such as Csapo, Wolfs, Von Leeuwen and Young (Csapo, 1970; Wolfs and van Leeuwen, 1979; Young, 2016).

Uterine activity triggers synchronous and coordinated uterine contractions that raise an intrauterine pressure, causing cervical softening and rupture of membranes. Spontaneous depolarization of myometrial cells generates a synchronous depolarization signal with an increase in connection proteins such as connexin; these junctions allow the flow of calcium ions and molecules smaller than 1000 Dalton through adjacent myocytes. These events cause a decrease in resting membrane potential, increasing its sensitivity to depolarization. Myocytes contain ion channels responsible for generating the action potential, including calcium and potassium channels that depend on voltage. The onset site of uterine activity, called pacemaker region, must be able to initiate a depolarization stream propagated in descending form that is transmitted to adjacent cells to cross the action potential (Young, 2018). Cells that are morphologically and immune-histochemically distinct from myocytes have been identified, called Cajal-like cells, which make up to 6% of myometrial cells. Recently, T-type calcium channels have been identified, however their function and participation in the uterine pacemaker system are still under investigation (Ciontea et al., 2005; Blanks et al., 2007; Young and Zhang, 2005).

Ion Channels

Potassium channels: They are responsible for keeping the cell membrane at rest, repolarizing the membrane to prevent the L-type calcium channels from opening. The voltage-dependent potassium channels are rapidly inactivated; normally their activation contributes to the uterine basal tone.

Calcium channels: They are the main source of electrical current entering the myometrium; they were classified as type L based on nifedipine sensitivity and their activation threshold is 55 to 50 mV (Figure 1).

Chloride channels: The opening of these channels in the myometrium generates a chloride flow, depolarizing the membrane. The activation of calcium chloride channels is triggered by the entry -and probably- the release of calcium from the rough endoplasmic reticulum. Recently, the anoctamin-1 (ANO1), also known as transmembrane member 16A (TMEM16A), has been described as a voltage-gated calcium-activated anion channel, which functions as a calcium-activated chloride channel (Yang et al., 2008). Selective antagonism of these channels in myometrial cells promotes the relaxation of spontaneous contractions (Wray et al., 2014).

The smooth muscle fibers undergo substantial increases in length, namely stretching, leading to an increase of the expression of oxytocin receptors, cytokines and gap junctions, as well as activation of the different ion channels. By joining oxytocin to its receptor in the myometrium, the PLC and phosphatidylinositol bisphosphate (IPP2) pathway are triggered, forming inositol 1,4,5-triphosphate (IP3), which acts as second messenger for the release of calcium in sarcoplasmic reticulum (SR). The latter triggers the recruitment and grouping of the sodium-myoinositol cotransporter-1 (SMIT1) transmembrane protein that forms the channel through which calcium and sodium enter, causing membrane depolarization, activating the voltage-dependent calcium channels and increasing the income of calcium into the cell (Figure 2).

Calcium binds to calmodulin to activate the myosin light chain kinase that phosphorylates myosin and allows the crossing of myosin with actin, causing myometrium contraction.

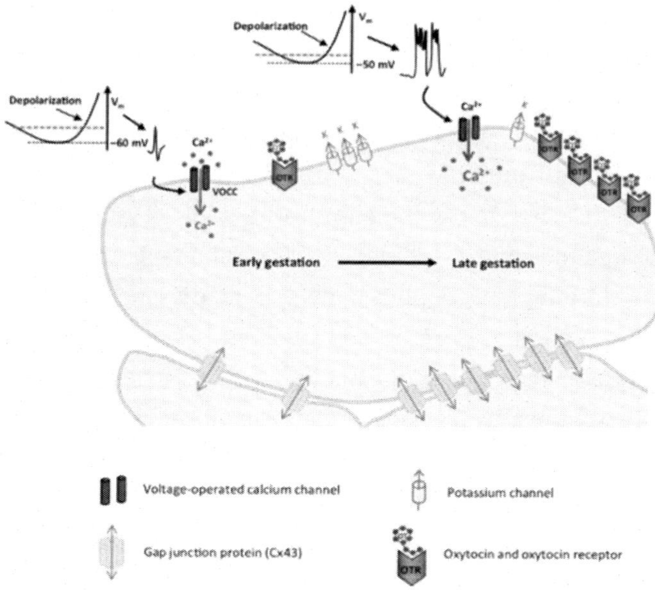

Figure 1. The change of ion channel currents during gestation. The oxytocin and oxytocin receptors are shown, playing an important role in the induction of uterine contractions during parturition.

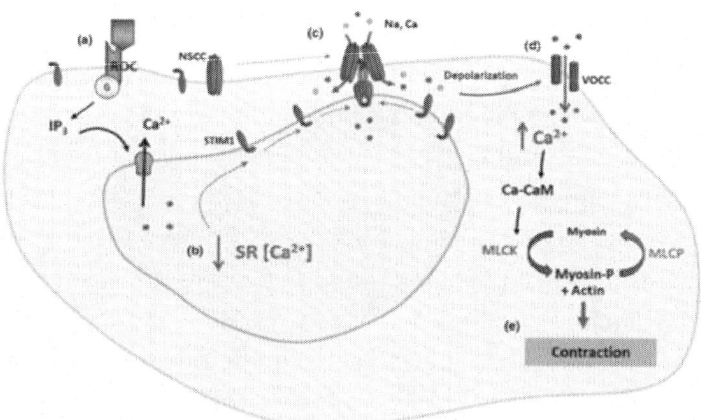

Figure 2. Calcium binds to calmodulin (CaM) to activate the myosin light chain kinase (MLCK) that phosphorylates myosin and allows the crossing of myosin with actin, causing myometrium contraction.

VITAMIN D AND ITS IMPLICATION IN UTERINE CONTRACTILITY

Vitamin D is the only nutrient that can be synthesized endogenously and works as a hormone. During pregnancy it plays a role in the anti-inflammatory and antimicrobial response within the fetoplacental unit. During pregnancy, the levels of $1,25(OH)_2D$ increase progressively from the first trimester. This increase is mainly due to the activity of 1-α hydroxylase in the maternal kidney, fetal, placenta, decidua and an increase in the binding protein (Adams and Hewison, 2012; Gray et al., 1979).

The endometrial decidua synthesizes $1,25(OH)_2D$ and $24,25(OH)_2D$, while the placenta synthesizes $24,25(OH)_2D$ (Weisman et al., 1979). Active vitamin D binds to vitamin D receptor (VDR) by unlocking the genes that synthesize antimicrobial peptides (Gombart, 2009), particularly the expression of cathelicidin in urogenital epithelial cells, in the monocyte macrophage system, in deciduous cells and in trophoblasts. Vitamin D has a well-established role in immunity, and it has been shown that its deficiency during pregnancy is related to an increase in maternal proinflammatory cytokines. The VDR is found in most immune system cells, including CD4+ and CD8+ T cells, which is why it is thought that this vitamin can act as an immunomodulator (Goldsmith 2015; Sassi et al., 2018; Szymczak and Pawliczak, 2015). T cells (thymus cells), B cells (bone marrow) and antigen-presenting cell (APC) are capable of synthesizing the active metabolite of vitamin D by expression of 1-α hydroxylase. This metabolite is involved in the regulation of the secretion of proinflammatory cytokines IL 4, IL 5, IL 10 and IL 13, and decreases the production of IL 6. Deficiency of $1,25(OH)_2D$ causes a decrease in cytokines anti-inflammatory and increased levels of IL-2, INF γ and TNF α (Thota et al., 2014).

Antimicrobial proteins that are synthesized by vitamin D have an action mainly against pathogens such as Listeria and group B Streptococci (Singh et al., 2015). $1,25(OH)_2D$ is related to a significant decrease in LPS-induced inflammation, resulting in a decrease of NF-κB expression, as vitamin D binds to its receptor increasing DNA synthesis and the expression of IκBα (Momentti et al., 2018).

Clinical Studies in Relation to Vitamin D and Uterine Contractility

Thota et al. (2014) evaluated the paracrine effects of monocytes and their regulation by vitamin D demonstrating that treatment with vitamin D decreased the mRNA expression of IL 1, 6, 13, TNFα, connexin 43, the oxytocin receptor and the NF-κB p65 in the nuclear fraction, decreasing the synthesis of cytokines in decidual NK cells. These results suggest that vitamin D decreases inflammatory markers induced by inflammation and contractile factors associated with the NF-κB pathway involved in the occurrence of preterm labor, due to infection and/or chronic inflammation (Thota et al., 2014).

Singh et al. (2015) showed that supplementation of 2000 international units (IU) of vitamin D in pregnant women from 12-16 weeks of gestation until pregnancy resolution reached sufficient levels of vitamin D (> 50 nmol/L) (Singh et al., 2015).

Yang et al. (2016) evaluated 200 pregnant women, comparing 46 preterm birth women with term pregnancy of the same age, finding no statistically significant differences for vitamin D levels with the risk of preterm birth (Yang et al., 2016).

The recommended dose for pregnant women of vitamin D is 600 IU; however, this recommendation is not sufficient to maintain serum levels of 25(OH)D above 20 ng/ml during pregnancy. Wagner et al. (2016) showed that levels of 25(OH)D > 40 ng/ml significantly reduced the risk of preterm birth so that a daily dose up to 4000 IU of vitamin D3 increases serum concentrations of vitamin D, normalizing the metabolism and homeostasis of calcium and vitamin D in pregnant women (Momentti et al., 2018; Wagner et al., 2016). Serum 25(OH)D is found significantly higher in women achieving a livebirth compared to those who did not. Indeed, supplementation of Vitamin D increases 25(OH)D levels raising the likelihood of a livebirth by 2% (Pal et al. 2016). A positive correlation between vitamin D levels in the serum and in the follicular fluids was found in a group of 84 patients, and a positive tendency to achieve clinical after *in vitro* fertilization (Ozkan et al. 2010).

MAGNESIUM SULPHATE

Magnesium is a trace element that plays a crucial role in more than 200 chemical reactions in the body. The ability of magnesium to prolong preterm birth has been described since more than 50 years ago. Clinical evidence shows a significant reduction in serum magnesium levels in preterm labor women (Okunade et al., 2014); a reduced permeability to calcium and magnesium regarding the exchange with fetal membranes was also observed (Lemancewicz et al., 2000). However, to date, the mechanism responsible for its tocolytic action is not yet fully characterized. One hypothesis states that magnesium competes with calcium to enter myometrial cells through the voltage-activated channels. Another possible explanation is that magnesium improves the activity of nitric oxide synthase (NOS3) (Parente et al., 2014). *In vitro* studies have shown that magnesium inhibits the entry of extracellular calcium, the release of intracellular calcium, the oscillations of cytosolic calcium and the physical contractions of smooth muscle in the myometrium (Popper et al., 1989; Phillippe, 1998). On the other hand, magnesium sulfate has been shown to inhibit the activity of PLC and the release of calcium in myometrial cells. It is one of the preferred approach to inhibit uterine activity for the risk of premature delivery, primarily because of its initial effect and minimal adverse effects. Several organic and inorganic magnesium salts can be administered orally and are used in various food supplements. It is well known that the bioavailability of magnesium is positively influenced by the concomitant intake of vitamin B6. In fact, they seem to interact together, and it has been suggested that vitamin B6 may improve the transport and accumulation of magnesium in the cells.

Normally, the required serum magnesium sulfate doses for effective tocolysis are 4-6 mEq/L (2-5 mmol/L) (Formin et al., 2006). Magnesium sulfate has an inhibitory effect on calcium channels activated by oxytocin, inhibiting the entry of intra and extracellular calcium; this leads to an increase in the activity of nitric oxide synthase (NOS) and an inhibition of the activity of specific phospholipase C phosphatidylinositol 4, 5 bisphosphate (Parente et al., 2014; Formin et al., 2006).

In non-pregnant women, a concentration of 12 to 14 mmol/L of magnesium sulfate inhibits spontaneous contractions and calcium absorption in the myometrium in few minutes, however, in the myometrium of pregnant women, lower concentrations (about 10 mmol/L) inhibit the flow of calcium mediated by oxytocin in 20 minutes.

Clinical Studies in Relation to Magnesium and Uterine Contractility

Victor et al. (2006) evaluated the effects of magnesium sulfate on contractile strength and the increase of free intracellular calcium in the myometrium of pregnant women, finding that at 5 mmol/L of magnesium sulfate the contractile response to calcium is inhibited and at 10 mmol/L it is immediately inhibited; however, the effects of magnesium sulfate did not reach statistical significance up to 20 minutes after exposure (Purisch and Gyamfi-Bannerman, 2017; Rabotti, and Mischi, 2014).

Parente et al. (2014) evaluated the efficacy of magnesium sulfate infusion for 48 hours as a second line tocolytic treatment in pregnant women between 22 and 34 weeks of gestation, whose uterine contractions were not sufficiently inhibited by betamimetics. The results of this study showed that after the infusion of magnesium sulfate, 90% of the patients prolonged their pregnancy for more than 48 hours. Interestingly, there were no maternal and neonatal side effects (Parente et al., 2014).

Alpha Lipoic Acid

Alpha lipoic acid is an organic sulfur compound, derived from octanoic acid, considered a natural molecule that exerts pleiotropic actions. It can exist in the R and S-enantiomeric forms. It is an essential cofactor of mitochondrial enzyme complexes important for aerobic metabolism. It is available in almost all foods, but in a very limited amount. It is found in vegetables such as potatoes, broccoli, spinach, tomatoes, brussels sprouts, peas and brown rice, which contain large amounts of this compound; however, red meat (especially liver, heart and kidney) is the most important source. Its production in humans is very low and the tissues that produce it are characterized by the presence of abundant mitochondria. It has been shown that alpha lipoic acid exerts a large number of immunomodulatory activities, mainly in preterm birth, counteracting inflammatory factors during pregnancy, through the following mechanisms:

- Reduction of the expression of the matrix of metalloproteinases (MMP) 9 and 16 induced by Tumor Necrosis Factor (TNF).
- In *in vitro* studies, alpha lipoic acid decreases the secretion of pro-inflammatory cytokines, such as TNF-α, IL-1β, IL-6, IL-8 and IL-20, induced by stimulation of LPS in mesangial cells.
- Effect on the inhibition of prostaglandin E2 (PGE2) mediated by the activity of cyclooxygenase-2 (COX-2).
- Activation of inducible nitric oxide synthase (iNOS) by alpha lipoic acid.
- Improvement of IL-10 levels.
- Inhibition of NF-κB (Li et al. 2015), a protein complex that controls the transcription of DNA that is activated by cytokines and thrombin as part of the induction of MMP, weakening fetal membranes. The inhibition of NF-kB translocation from the cytoplasm to the nucleus has a pivotal role in reducing uncontrolled oxidative and inflammatory reactions (Monastra et al. 2016).
- Oxygen free radical scavenging (because it is an antioxidant compound), promotes the regeneration of endogenous antioxidants such as vitamin C and E, exerts anti-inflammatory properties and interacts synergistically with magnesium.

Clinical Studies in Relation to Magnesium and Uterine Contractility

Facchinetti et al. (2017) studied women at risk of preterm labor, comparing the effects of vaginal alpha lipoic acid (400 mg/day) with placebo. They showed that supplementation with alpha lipoic acid during the second trimester proved to be effective in reducing symptoms, as well as in preventing both the onset of symptoms and shortening of the cervix in women at risk of premature delivery. Reports from previous studies suggest that the administration of alpha lipoic acid is safe in therapeutic doses in pregnant women, without any adverse effects (Grandi et al, 2017).

- Vitrano et al. (2018) conducted a study where alpha lipoic acid was administered orally (300 mg, twice daily for 30 days) and vaginally (10 mg, once daily for 10 days). The patients were evaluated at the beginning of the treatment, at 7 and 30 days, and at 34 weeks of gestation, where variables such as maternal characteristics, symptoms and cervical length were evaluated. The administration of combined alpha lipoic acid showed efficacy in reducing symptoms and preventing cervical shortening.
- Porcaro et al. (2015) conducted a randomized controlled clinical trial of pregnant women with miscarriage. The study aimed to observe the effects of alpha lipoic acid supplementation (600 mg orally) in the treatment of miscarriage compared to the standard approaches such as vaginal progesterone suppositories, observing its effects in the healing of subcorionic hematomas and also in the reduction of subjective and objective signs like vaginal bleeding, abdominal pain and uterine contractions. Patients suffered from pelvic pain with or without moderate vaginal bleeding and in all subjects the presence of subcorionic hematoma was diagnosed. The follow-up of the main signs of the miscarriage (separation of the chorioamniotic membranes and subcorionic hematoma, vaginal bleeding, abdominal pain, and uterine contractions) decreased and

disappeared in the group treated with alpha lipoic acid faster than in the group treated only with progesterone. The clinical evolution of the healing of chorioamniotic membranes and the resorption of the subcorionic hematoma was clearly greater in patients treated with alpha lipoic acid, observing a greater improvement compared to the group that received progesterone.

Alpha lipoic acid has been shown to act as an immunomodulatory molecule in different inflammatory diseases, where selective and final regulation of the intertwined pathways must be established. It is also able to modulate the complex of cells and molecules that participate in the miscarriage, offering new ways to substantially reduce complications linked to the miscarriage, and improve the outcome of pregnancy (Monastra et al., 2016; Vitrano et al., 2018;).

Conclusion

In conclusion, these molecules, vitamin D, magnesium sulfate and alpha lipoic acid, exerting different actions are able to counteract the premature uterine contractility and improve the outcome of pregnancy. Therefore, pregnant women at risk of preterm birth, or encountering unexpectedly premature uterine contractions, should take in consideration supplementation of vitamin D, magnesium sulfate and alpha lipoic acid. The supplementation of these 3 molecules might work synergistically and modulate the erroneous uterine activity, that could cause serious complications for the fetus and the mother.

References

Adams, J. S. & Hewison, M. (2012). "Extrarenal Expression of the 25-Hydroxyvitamin D-1-Hydroxylase". *Arch Biochem Biophys.*, *1*, 523(1), 95–102. doi: 10.1016/j.abb.2012.02.016.

Alvarez, H. & Caldeyro-Barcia, R. (1954). "The normal and abnormal contractile waves of the uterus during labour". *Gynaecologia*, *138*(2), p. 190-212.

American College of Obstetricians and Gynecologists. (2013). "ACOG committee opinion no. 561: nonmedically indicated early-term deliveries." *Obstet Gynecol.*, *121*(4), 911–915.

Blanks, A. M., Zhao, Z. H., Shmygol, A., Bru-Mercier, G., Astle, S. & Thornton, S. (1970). "Characterization of the molecular and electrophysiological properties of the T-type calcium channel in human myometrium". *J Physiol*, 2007, *581*(Pt 3), p. 915-26.

Carson, R. (2004). "Detection and Prevention of Premature Labour." *Neuro Endocrinol Lett*, *25*, 35–41.

Ciontea, S. M., Radu, E., Regalia, T., Ceafalan, L., Cretoiu, D., Gherghiceanu, M., Braga, R. I., Malincenco, M., Zagrean, L., Hinescu, M. E. & Popescu, L. M. (2005). "C-kit immunopositive interstitial cells (Cajal-type) in human myometrium". *J Cell Mol Med*, *9*(2), p. 407-20.

Committee on Practice Bulletins—Obstetrics. (2012). "The American College of Obstetricians and Gynecologists. Practice bulletin no. 130: prediction and prevention of preterm delivery." *Obstet Gynecol.*, *120*(4), 964–973.

Csapo, A. (1970). "The diagnostic significance of the intrauterine pressure". *I. Obstet Gynecol Surv*, *25*(5), p. 403-35.

Fomin, V. P., Gibbs, S. G., Vanam, R., Morimiya, A. & Hurd, W. W. (2006). "Effect of magnesium sulfate on contractile force and intracellular calcium concentration in pregnant human myometrium". *Am J Obstet Gynecol.*, *194*, 1384–90.

Goldsmith, J. (2015). "Vitamin D as an Immunomodulator: Risks with Deficiencies and Benefits of Supplementation". *Healthcare (Basel)*, *3*(2), 219–232. doi: 10.3390/healthcare3020219.

Gombart, A. (2009). "The vitamin D–antimicrobial peptide pathway and its role in protection against infection". *Future Microbiol.*, *4*, 1151.

Grandi, G., Pignatti, L., Ferrari, F., Dante, G., Neri, I. & Facchinetti, F. (2017). "Vaginal alpha-lipoic acid shows an anti-inflammatory effect on the cervix, preventing its shortening after primary tocolysis. A pilot, randomized, placebo-controlled study". *J Matern Fetal Neonatal Med*, *30*, 2243-2249.

Gray, T. K., Lester, G. E. & Lorenc, R. S. (1979). "Evidence for extra-renal 1 alpha-hydroxylation of 25-hydroxyvitamin D3 in pregnancy". *Science*, *22*, 204(4399), 1311-1313.

Griffin, J. B., Jobeì, A. H., Rouse, D., McClure, E. M., Goldenberg, R. L. & Kamath-Rayne, B. D. (2019). "Evaluating WHO-Recommended Interventions for Preterm Birth: A Mathematical Model of the Potential Reduction of Preterm Mortality in Sub-Saharan Africa". *Glob Health Sci Pract.*, *24*, 7(2), 215–227. doi:10.9745/GHSP-D-18-00402.

Groom, K. (2007). "Preterm Labour. Pharmacological Prevention of Prematurity." *Best Pract Res Clin Obstet Gynaecol*, *21*, 843–856.

Hubinont, C. & Debieve, F. (2011). "Prevention of Preterm Labour: Update on Tocolysis." *J Pregnancy*, 941–57.

Lemancewicz, A., Laudańska, H., Laudański, T., Karpiuk, A. & Batra, S. (2000). "Permeability of fetal membranes to calcium and magnesium: possible role in preterm labour". *Hum Reprod.*, *15*(9), 2018-2022. Doi:10.1093/humrep/15.9.2018.

Li, G., Fu, J., Zhao, Y., Ji, K., Luan, T. & Zang, B. (2015). "Alpha-lipoic acid exerts anti-inflammatory effects on lipopolysaccharide-stimulated rat mesangial cells via inhibition of nuclear factor kappa B (NF-kappaB) signaling pathway." *Inflammation*, *38* (2), 510-9. doi: 10.1007/s10753-014-9957-3.

Momentti, A. C., Estadella, D. & Pellegrini Pisani, L. (2018). "Role of vitamin D in pregnancy and Toll-like receptor pathway". *Steroids*, *137*, 22–29. doi:10.1016/j.steroids.2018.07.009.

Monastra, G., De Grazia, S., Cilaker Micili, S., Goker, A. & Unfer, V. (2016). "Immunomodulatory activities of alpha lipoic acid with a special focus on its efficacy in preventing miscarriage". *Expert opinion on drug delivery*, *13*(12), 1695-1708.

Okunade, K. S., Oluwole, A. A. & Adegbesan-Omilabu, M. A. (2014). A Study on the Association between Low Maternal Serum Magnesium Level and Preterm Labour. *Advances in medicine*, *704875*. doi:10.1155/2014/704875.

Oshiro, B. T., Kowalewski, L., Sappenfield, W., et al. (2013). "A multistate quality improvement program to decrease elective deliveries before 39 weeks of gestation" [published correction appears in *Obstet Gynecol.*, *122*(1), 160]. *Obstet Gynecol.*, *121*(5), 1025–1031.

Ozkan, S., Jindal, S., Greenseid, K., et al. (2010). "Replete vitamin D stores predict reproductive success following *in vitro* fertilization". *Fertil Steril.*, *94*, 1314–1319.

Pal, L., Zhang. H., Williams. J., et al. (2016). "Reproductive Medicine Network. Vitamin D status relates to reproductive outcome in women with polycystic ovary syndrome: Secondary analysis of a multicenter randomized controlled trial". *J. Clin. Endocrinol. Metab.*, *101*, 3027-3035.

Parente, E., Colannino, G. & Ferrara, P. (2014). "Efficacy of magnesium and alpha lipoic acid supplementation in reducing premature uterine contractions". *Open Journal of Obstetrics and Gynecology*, *4*(09), 578.

Parente, E., Colannino, G. & Ferrara, P. (2014). "Efficacy of Magnesium and Alpha Lipoic Acid Supplementation in Reducing Premature Uterine Contractions". *Open Journal of Obstetrics and Gynecology*, *4*, 578-583.

Phillippe, M. (1998). "Cellular mechanisms underlying magnesium sulfate inhibition of phasic myometrial contractions". *Biochem Biophys Res Commun.*, *252*(2), 502-7.

Popper, L. D., Batra, S. C. & Akerlund, M. (1989). "The effect of magnesium on calcium uptake and contractility in the human myometrium". *Gynecol Obstet Invest.*, *28*(2), 78-81.

Porcaro, G., Brillo, E., Giardina, I. & Di Iorio, R. (2015). "Alpha Lipoic Acid (ALA) effects on subchorionic hematoma: preliminary clinical results". *Eur Rev Med Pharmacol Sci*, *19*, 3426-32.

Purisch, S. E. & Gyamfi-Bannerman, C. (2017). Epidemiology of preterm birth. In *Seminars in perinatology*, (Vol. *41*, No. 7, pp. 387-391). WB Saunders.

Rabotti, C. & Mischi, M. (2014). Propagation of electrical activity in uterine muscle during pregnancy: a review. *Acta Physiologica*, *213*(2), 406–416. doi:10.1111/apha.12424.

Rood, K. M. & Buhimschi, C. S. (2017). Genetics, hormonal influences, and preterm birth. *Seminars in Perinatology*, *41*(7), 401–408. doi:10.1053/j.semperi.2017.07.011.

Sassi, F., Tamone, C. & D'Amelio, P. (2018). "Vitamin D: Nutrient, Hormone, and Immunomodulator". *Nutrients*., Nov, *10*(11), 1656. doi: 10.3390/nu10111656.

Singh, J., Hariharan, C. & Bhaumik, D. (2015). "Role of vitamin D in reducing the risk of preterm labour, *International Journal of Reproduction, Contraception, Obstetrics and Gynecology Int J Reprod Contracept Obstet Gynecol.*, *4*(1), 86-93.

Szymczak, I. & Pawliczak, R. (2016). "The Active Metabolite of Vitamin D3 as a Potential Immunomodulator". *Scandinavian Journal of Immunology*, *83*(2), 83-91. doi.org/10.1111/sji.12403.

Thota, C., Laknaur, A., Farmer, T., Ladson, G., Al-Hendy, A. & Ismail, N. (2014). "Vitamin D regulates contractile profile in human uterine myometrial cells via NF-kB pathway". *Am J Obstet Gynecol.*, *210*(4), 347, e1-347.e10. doi: 10.1016/j.ajog.2013.11.027.

Vitrano, G., Mocera, G., Guardino, M., Giallombardo, V. & Venezia, R. (2018). Oral plus vaginal alpha-lipoic acid in women at risk for preterm delivery. *IJMDAT*, *1*(1), 104.

Vogel, J. P., Chawanpaiboon, S., Moller, A. B., Watananirun, K., Bonet, M. & Lumbiganon, P. (2018). The global epidemiology of preterm birth. *Best practice & research Clinical obstetrics & gynaecology*, *52*, 3-12.

Wagner, C. L., Baggerly, C., McDonnell, S., Baggerly, K. A., French, C. B., Baggerly, L., Hamilton, S. A. & Hollis, B. W. (2016). "Post-hoc analysis of vitamin D status and reduced risk of preterm birth in two vitamin D pregnancy cohorts compared with South Carolina March of Dimes 2009–2011 rates". *J. Steroid Biochem. Mol. Biol.*, *155*, 245–251. https://doi.org/10.1016/j.jsbmb.2015.10.022.

Wei, S. Q., Fraser, W. & Luo, Z. C. (2010). "Inflammatory cytokines and spontaneous preterm birth in asymptomatic women: a systematic review." *Obstetrics and Gynecology*, *116*, (2 Pt 1), 393-401. doi:10.1097/aog.0b013e3181e6dbc0.

Weisman, Y., Harell, A., Edelstein, S., David, M., Spirer, Z. & Golander, A. (1979). "1 alpha, 25-Dihydroxyvitamin D3 and 24,25-dihydroxyvitamin D3 *in vitro* synthesis by human decidua and placenta". *Nature*, *281*, 317–9.

Wolfs, G. M. & van Leeuwen, M. (1979). "Electromyographic observations on the human uterus during labour". *Acta Obstet Gynecol Scand Suppl*, *90*, p. 1-61.

Wray, S., Burdyga, T., Noble, D., Noble, K., Borysova, L. & Arrowsmith, S. (2014). Progress in understanding electro-mechanical signalling in the myometrium. *Acta Physiologica*, *213*(2), 417–431. doi:10.1111/apha.12431.

Yang, L., Pan, S., Zhou, Y., Wang, X., Qin, A., Huang, Y. & Sun, S. (2016). The Correlation Between Serum Vitamin D Deficiency and Preterm Birth. *Medical Science Monitor*, *22*, 4401–4405. doi:10.12659/ msm. 898117.

Yang, Y. D., Cho, H., Koo, J. Y., Tak, M. H., Cho, Y., Shim, W. S., Park, S. P., Lee, J., Lee, B., Kim, B. M., Raouf, R., Shin, Y. K. & Oh, U. (2008). "TMEM16A confers receptor-activated calcium-dependent chloride conductance". *Nature.*, *455* (7217), 1210–5. doi:10.1038/nature07313.

Young, R. C. & Zhang, P. (2005). Inhibition of *in vitro* contractions of human myometrium by mibefradil, a T-type calcium channel blocker: support for a model using excitation- contraction coupling, and autocrine and paracrine signaling mechanisms. *J Soc Gynecol Investig.*, *12*(4), p. e7-12.

Young, R. C. (2016). Mechanotransduction mechanisms for coordinating uterine contractions in human labor. *Reproduction*, *152*(2), p. R51-61.

Young, R. C. (2018). The uterine pacemaker of labor. *Best Practice & Research Clinical Obstetrics & Gynaecology.* doi:10.1016/ j.bpobgyn. 2018.04.002.

Zhou, S. S., Tao, Y. H., Huang, K., Zhu, B. B. & Tao, F. B. (2017). "Vitamin D and risk of preterm birth: Up-to-date meta-analysis of randomized controlled trials and observational studies". *Journal of Obstetrics and Gynaecology Research*, *43*(2), 247–256. doi:10.1111/ jog.13239.

In: Vitamin D as Progesterone-Like Hormone ISBN: 978-1-53617-516-5
Editor: Vittorio Unfer

Chapter 6

VITAMIN D IN ASSISTED REPRODUCTIVE TECHNOLOGY (ART): THE THERAPEUTIC REGIMEN

Stamatios Petousis[1,2,*], MD, MSc, PhD,
Konstantinos Ravanos[1], MD, MSc,
Chrysoula Margioula-Siarkou[1], MD, MSc, PhD,
George Mavromatidis[2], MD, PhD,
Konstantinos Dinas[2], MD, MSc, PhD,
Yannis Prapas[1], MD, PhD
and Nikolaos Prapas[1], MD, MSc, PhD
[1]IAKENTRO, Infertility Treatment Center,
Thessaloniki, Greece
[2]2nd Department of Obstetrics and Gynaecology,
Aristotle University of Thessaloniki,
Thessaloniki, Greece

[*] Corresponding Author's Email: petousisstamatios@gmail.com.

ABSTRACT

Vitamin D has recently drawn attention for many aspects of women's health. In particular, increasing evidence suggests a link between vitamin D and current gynaecologic and fertility concerns. As modern Reproductive Technologies mainly aim to provide treatments tailored to each couple's specific needs, supplementation with vitamins and administration regimens are now included in the therapeutic strategy. Given that women undergoing *in-vitro* fertilization (IVF) generally receive multiple medications, physicians must define the optimal therapeutic regimen of vitamin D supplementation in order to improve compliance. The aim of the present review is to summarize the available evidence regarding the potential impact of vitamin D deficiency and supplementation on IVF outcomes. Special attention, in particular, will be devoted to vitamin D deficiency as etiological factor in female infertility of different causes.

Keywords: vitamin D, Assisted Reproductive Technology, infertility, PCOS, supplementation

INTRODUCTION

Vitamin D has recently become the center of interest for many aspects of women's health (Franasiak, Lara and Pellicer 2017). Despite a long history of being associated with bone metabolism, recent findings suggest other roles for vitamin D. Indeed, the etiopathogenesis of female malignancies, cardiovascular and metabolic diseases may directly derive or be strongly associated with vitamin D deficiency (Yoshizawa et al. 1997, Heaney 2008, Holick 2007).

Increasing evidence also suggests a link between vitamin D and current gynaecologic and fertility concerns (Flood-Nichols et al. 2015), supported by the presence of vitamin D receptors (VDRs) on the ovary, endometrium and placenta (Franasiak, Lara and Pellicer 2017). As a multifunctional molecule that affects all aspects of reproductive life, having a role even in pregnancy complications, vitamin D may be critically important both for fertility status and final outcome of Assisted Reproductive Technology

(ART) procedures (Gernand et al. 2014, Christesen et al. 2012). In this context, the potential impact of vitamin D deficiency, and thereafter supplementation, has lately drawn a great deal of interest.

As modern ART mainly aims to provide tailored treatments for each couple's reproductive defects, supplementation with vitamins and administration regimens have been considered as part of the whole therapeutic strategy. Given that women undergoing *in-vitro* fertilization (IVF) generally receive multiple medications, physicians constantly have to evaluate the need and the optimal therapeutic regimen of vitamin D supplementation in order to improve compliance. Moreover, in-depth investigation on the exact role of vitamin D in IVF is highly desirable to avoid potentially negative interactions with the other medications.

Based on the data from the literature, whether physician should always include vitamin D supplementation in IVF treatment is still unclear, just as much as the administration timing, the dosage, and whether it should be adjusted to every patient's different clinical status (Blomberg Jensen 2014, Thomson, Spedding and Buckley 2012).

The present review summarizes the available evidence regarding the potential impact of vitamin D deficiency and supplementation on IVF outcomes. The studies reported are classified based on whether their findings support or question the role of vitamin D in IVF/ART final outcome, in order to draw conclusions according to current clinical evidence and point out issues that require further investigation.

Vitamin D Involvement in Physiology of Reproduction

The role of vitamin D in human reproduction is manifold. First of all, it has a potential effect on follicles and embryos. Knockout of VDR genes results in detrimental effects on folliculogenesis in mice (Yoshizawa et al. 1997). Moreover, vitamin D stimulates the production of Insulin-like Growth Factor Binding Protein-1 (IGFBP-1) and displays beneficial effects on ovarian steroidogenesis, being correlated with levels of estradiol, estrone, progesterone and anti-müllerian hormone (AMH) in adults (Parikh et al. 2010, Aflatoonian et al. 2010, Zarnani et al. 2010).

Vitamin D greatly affects endometrial receptivity. Indeed, VDR and the enzyme that activates vitamin D (1α-hydroxylase) are both expressed in the endometrium. Moreover, VDR and 1α-hydroxylase knockout mice exhibit an undeveloped uterus and are infertile (Lerchbaum and Obermayer-Pietsch 2012). Another study demonstrated that calcitriol ($1,25(OH)_2D_3$) can upregulate the expression of HOXA10, an essential protein for female fertility and embryo implantation, by binding to its receptor in the endometrial stromal cells (Vigano et al. 2006).

Recent researches demonstrated that vitamin D is important not only for embryo implantation but for gestation as well. Clinical studies proved that pregnant women with vitamin D deficiency are more likely to develop preeclampsia or gestational diabetes and to require cesarean section (Merewood et al. 2009, Shand et al. 2010, Soheilykhah et al. 2010). Both vitamin D and HOXA10 are expressed in the endometrium, decidua and placenta throughout the pregnancy (Evans et al. 2006), suggesting that vitamin D plays a crucial role in keeping healthy gestation (Bodnar et al. 2007).

As a conclusion, strong indications support the role of vitamin D in various aspects of reproduction, including conception, implantation, trophoblast invasion, gestation and maternal health status during pregnancy.

Studies Supporting Vitamin D Role in ART Outcome

Even though the literature reports heterogeneous findings, the majority of published studies supports the beneficial impact of vitamin D on ART outcome.

One of the strongest scientific evidence comes from the meta-analysis by Zhao et al. (Zhao et al. 2018), which included a total of 9 cohort studies aiming to provide evidence on live birth rate and both clinical and ongoing pregnancy. The results show a 26% decrease of live birth rate in women with vitamin D deficiency [RR 0.74, (95% CI 0.58 – 0.90)], along with a trend for lower clinical pregnancy rates, albeit non-significant. The authors conclude recommending the supplementation in women with vitamin D deficiency, as this is associated with lower probability of live birth after IVF/Intra-Cytoplasmic Sperm Injection (ICSI). Some flaws in this meta-analysis must be, however, taken into account: Six out of nine studies are retrospective, despite the authors claiming they are high-scored on the Newcastle-Ottawa Quality Assessment Scale; the included studies used different stimulation protocols; the population of infertile women was heterogenous, as the studies included both IVF and oocyte donation.

The conclusions of this meta-analysis are further supported by a more recent meta-analysis (Chu et al. 2018). The authors included 11 studies, aiming to assess whether vitamin D deficiency correlates with final IVF outcome. Live birth rate was 33% higher in women with replete vitamin D status, compared to insufficient or deficient women (OR 1.33 [1.08 - 1.65]). Five of the included studies (accounting for 1700 patients) also report that women with replete vitamin D status were 34% more likely to test positive for human chorionic gonadotropin (hCG) (OR 1.34 ([1.04 - 1.73]) and to achieve clinical pregnancy (OR 1.46 [1.05 - 2.02]). No association was found between vitamin D status and miscarriage risk.

Also this meta-analysis includes studies with heterogeneous patient cohorts and different protocols for fertility treatment. Furthermore, Iliuta et al. (Iliuta, Pijoan and Matorras 2018) pointed out a series of statistical and methodological issues in a *letter to the Editor* published in Human Reproduction Journal. They conclude that the results would be different with more precise data extraction and with a different methodological approach. However, they acknowledge the significant decrease of all ART outcomes in vitamin D deficient patients.

In light of the methodological flaws in these meta-analyses, and the potential bias given by the patient cohort in the included studies, a large randomized controlled trial (RCT) is warranted to confirm the beneficial role of vitamin D in IVF populations.

The study by Polyzos et al. (Polyzos et al. 2014) analyzes the impact of vitamin D deficiency on pregnancy rates in infertile couples undergoing IVF/ICSI. Despite being retrospective, it is one of the few carried out on a well-defined population of patients, namely those undergoing single embryo blastocyst transfer on day 5 after oocyte retrieval. Also, the authors defined precisely *deficiency* as serum levels of 25-hydroxyvitamin D (25(OH)D) lower than 20 ng/ml, in accordance with the clinical practice guidelines from the Institute of Medicine and the Endocrine Society. Logistic regression analysis indicates that vitamin D deficiency was independently associated with 44% reduction of clinical pregnancy rates when considering only women that undergo elective single embryo blastocyst transfer [OR (95% CI) 0.56 (0.33 - 0.93), P = 0.024]. The authors, anyway, advise to carry out further prospective studies to validate these results and to investigate the mechanistic details of how vitamin D levels impair pregnancy rates in infertile women undergoing IVF/ICSI.

The cause of infertility in patients undergoing ART procedures is an aspect to take into consideration. Impaired fertility due to ovarian or tubal factors, endometriosis, or to unknown reasons have totally different etiology. Relating the effect of vitamin D on the different causes of infertility is fundamental to define supplementation regimens in ART protocols.

Polycystic Ovary Syndrome

Vitamin D proved to have a beneficial impact on patients with Polycystic Ovary Syndrome (PCOS). Fatemi et al. investigated the role of supplementation with vitamin E and vitamin D_3 on ICSI outcomes in women with polycystic ovary syndrome (Fatemi et al. 2017). The RCT included a total of 105 infertile PCOS women scheduled to receive either a combination of vitamin E and vitamin D_3 (vitamin E, 400mg/day and vitamin D_3, 50,000 IU/two weeks, n = 52) or placebo (= 53) for 8 weeks after ICSI procedure. The primary outcomes were implantation rate, and both biochemical and clinical pregnancy rates. The secondary included oocyte and embryo quality, fertilization rate and variation in serum malondialdehyde (MDA), total antioxidant capacity (TAC) and vitamin D_3 after the treatment. At the end of the study, all primary outcomes significantly improved in the supplementation group with respect to placebo (biochemical pregnancy rate: 69% vs 25.8%; clinical pregnancy rate: 62.1% vs. 22.6%; implantation rate: 35.05% vs only 8.6%). Moreover, further analysis demostrated a positive association between vitamin D_3 levels and both implantation and clinical pregnancy rates. On the other hand, TAC does not increase in the treated group, suggesting that the positive effects of supplementation on ICSI outcomes are unrelated to the antioxidant activity, but rather to pathways not yet elucidated.

The prospective study by Zhao et al. evaluated the IVF/ICSI outcome in 305 infertile PCOS patients, according to their vitamin D status (Zhao et al. 2019).

The authors divided the patients into four groups: deficiency group without treatment, normal group without treatment, normal group after treatment and deficiency group after treatment. The observed clinical pregnancy rates, respectively of 19.3%, 65.2%, 66.7% and 23.5%, indicate that patients with vitamin D deficiency are about 3 times less likely to carry on a pregnancy following the IVF/ICSI procedure, and that supplementation significantly improve this outcome. In contrast with the finding of Fatemi et al. (Fatemi et al. 2017), the authors also observed particularly marked differences in the number of high-quality embryos and available embryos for transfer. They concluded that vitamin D supplementation restores serum levels of vitamin D in infertile women with PCOS and IR, leading to improved embryo quality and to significantly higher clinical pregnancy rates.

Dastorani et al. suggested that improved metabolic profile of lipids and insulin may account for the beneficial impact that vitamin D supplementation has on IVF/ICSI outcomes in PCOS women (Dastorani et al. 2018).

In this randomized, double-blinded, placebo-controlled trial, 40 infertile PCOS women were assigned to two groups to receive either vitamin D (50,000 IU) or placebo (n = 20 each group) every other week, for 8 weeks.

Gene expression for insulin and lipid metabolism was conducted using peripheral blood mononuclear cells (PBMCs), via RT-PCR method. The authors observed that vitamin D supplementation resulted in a significant reduction in serum AMH, insulin levels (-1.4 ± 1.6 vs. -0.3 ± 0.9μIU/mL, $p = 0.007$) and homeostatic model of assessment for insulin resistance, while significantly increasing the quantitative insulin sensitivity check index, compared with the placebo.

Moreover, both total and LDL-cholesterol serum levels decreased significantly in the study group.

Endometriosis

Although limited, the current literature suggests a strong etiopathogenic association between vitamin D and endometriosis.

A recently published prospective observational study (Ciavattini et al. 2017) includes 49 women of childbearing age, diagnosed with singleton ovarian endometrioma. The authors found that 85.7% of patients had hypovitaminosis D, and that a significant linear correlation existed between serum levels of vitamin D and the diameter of ovarian endometrioma. Miyashita et al. (Miyashita et al. 2016) tried to clarify *in vitro* the effects of vitamin D_3 on human endometrial cells, isolated from an ovarian endometrioma. Cell cultures with $1,25(OH)_2D_3$ display reduced inflammatory responses induced by IL-1β or TNF-α, suggesting that vitamin D modulates both inflammation and proliferation in endometrial cells. Even though vitamin D supplementation may represent a novel therapeutic strategy for women with endometriosis, the beneficial impact on IVF/ICSI outcomes in these patients still requires confirmation from prospective observational studies or RCTs.

Unexplained Infertility

Only two publications appear in Pubmed search using the keywords "vitamin D and unexplained infertility".

Djurovic et al. studied the DNA from 117 female patients with unexplained infertility and 130 fertile controls in order to check for VDR genotypes (Fokl, Bsml, Apal and Taql) using PCR-RFLP (Djurovic et al. 2018).

They demonstrated a significant association of Fokl and Bsml polymorphisms with infertility, concluding that alterations in the gene expression of VDR may represent an etiological factor for unexplained infertility.

In a retrospective cohort study, Butts et al. evaluated the relationship between vitamin D deficiency and reproductive outcomes after ovarian stimulation in women either with PCOS or unexplained infertility (Butts et al. 2019).

The authors pointed out the significant association of vitamin D levels with ovarian response in PCOS patients, but failed to find a similar association in patients with unexplained infertility.

Tubal Factor Infertility (TFI)

The prospective observational study by Chen et al. investigates the relationship between 25(OH)D and female infertility of various causes (Chen et al. 2018).

After adjusting for confounders, they found a dose-effect relation between IL-6 tertiles and TFI: the higher the IL-6, the higher the risk for TFI.

IL-6, on the other hand, proved to negatively correlate with 25(OH)D. Despite both vitamin D deficiency and high serum IL-6 level are risk factors for TFI, the literature lacks studies regarding the potential beneficial impact of vitamin D supplementation in IVF cycles exclusively for women with such condition.

Studies Questioning Vitamin D Role on ART Outcome

Despite growing evidence supports the beneficial role of vitamin D supplementation on IVF/ICSI outcomes, there are a few studies that fail to find a significant effect of vitamin D levels on reproduction.

The restrospective investigation by Jiang et al. aims to assess the effect of serum vitamin D levels on embryo development and clinical outcomes in couples undergoing IVF/ICSI (Jiang et al. 2019).

To our knowledge, the study included the largest number of patients reported in the literature, with 1,883 women and 1,720 men. Female patients were divided into four groups, according to their serum 25(OH)D quartiles.

The authors observed no statistically significant difference regarding top quality embryos both on day 2 and on day 5/6. Likewise, they report no significant difference or trend for implantation, clinical pregnancy, abortion and live birth rates.

It should be mentioned, however, that the authors did not investigate the potential effect of vitamin D supplementation on IVF/ICSI outcome. Moreover, the study population encompasses only patients from East Asia, without considering genetic alterations with respect to other ethnicities.

Banker et al. examined the effect of vitamin D deficiency in both oocyte donors and their recipients (Banker, Sorathiya and Shah 2017). Their prospective trial included a total of 192 participants, with both donors and recipients classified as vitamin D deficient or replete-insufficient.

The authors found no significant difference in the endometrial thickness, the number of oocytes retrieved or embryos transferred, the dose of gonadotropins, the percentage of mature oocytes, the percentage of usable embryos, the fertilization and the clinical pregnancy rate. Just like the previos study, the role of vitamin D supplementation was completely neglected. Moreover, patients with replete vitamin D status and controls for covariates were not included.

Critical Approach, Final Conclusions and Proposals

Current evidence on the role of vitamin D in ART leads to two considerations: *1)* strong indications support the beneficial effects of supplementation on IVF/ICSI outcomes; *2)* more RCTs, with appropriate investigation methodology, are required to better define the therapeutic regimen of such supplementation and its role in treating infertility with different etiology.

Noteworthy, vitamin D seems to significantly improve the fertility of PCOS women, probably due to its involvement in the metabolism of insulin and lipids. Much fewer data indicate an effect of vitamin D on infertility caused by endometriosis or tubal factors. Reports, in this regards, are mainly observational and do not investigate the impact of vitamin D supplementation on IVF/ICSI outcomes. The effect of vitamin D supplementation in patients with unexplained infertility remains largely underexplored and requires further investigation with properly designed prospective trials.

In conclusion:

- the beneficial role of vitamin D in ART outcomes is supported by strong indications reported in the literature;
- vitamin D supplementation seems to significantly improve the fertility in PCOS women;
- further evidence is required to assess the effect of vitamin D supplementation in patients with unexplained infertility;
- prospective RCTs are warranted for each subcategory of infertile patients;
- further attention should be directed towards the exact role of vitamin D on patients undergoing oocyte donation cycles.

Ultimately, supplementation with vitamin D to patients undergoing ART procedures has evident positive impact on the reproductive outcomes. Specific effect, regimen and duration of the treatment are still debated, and further evidence is required to tailor the therapy to the patients' specific needs. However, based on the retrieved data, supplementation of vitamin D should be encouraged in deficient women undergoing IVF. Due to its progesterone-like hormone activity, supplementation should specifically be given during the luteal phase of the ovarian cycle, and not in the follicular phase, continuing until the 12th week of gestation, if pregnancy occurs.

REFERENCES

Aflatoonian, A., M. Mashayekhy, F. Mohamadian and F. M. Moghaddam. 2010. "The correlation between follicular fluid anti-mullerian hormone levels and fertilization and embryo quality in ART cycles". *International Journal of Reproductive BioMedicine,* 8 (5):157 - 0.

Banker, M., D. Sorathiya and S. Shah. 2017. "Vitamin D Deficiency Does Not Influence Reproductive Outcomes of IVF-ICSI: A Study of Oocyte Donors and Recipients". *J. Hum. Reprod. Sci.,* 10 (2):79 - 85. doi: 10.4103/jhrs.JHRS_117_16.

Blomberg, J. M. 2014. "Vitamin D and male reproduction". *Nat. Rev. Endocrinol.,* 10 (3):175 - 86. doi: 10.1038/nrendo.2013.262.

Bodnar, L. M., J. M. Catov, H. N. Simhan, M. F. Holick, R. W. Powers and J. M. Roberts. 2007. "Maternal vitamin D deficiency increases the risk of preeclampsia". *J. Clin. Endocrinol. Metab.,* 92 (9):3517 - 22. doi: 10.1210/jc.2007-0718.

Butts, S. F., D. B. Seifer, N. Koelper, S. Senapati, M. D. Sammel, A. N. Hoofnagle, A. Kelly, S. A. Krawetz, N. Santoro, H. Zhang, M. P. Diamond and R. S. Legro. 2019. "Vitamin D Deficiency Is Associated With Poor Ovarian Stimulation Outcome in PCOS but Not Unexplained Infertility". *J. Clin. Endocrinol. Metab.,* 104 (2):369 - 378. doi: 10.1210/jc.2018-00750.

Chen, W., X. Jiao, J. Zhang, L. Wang and X. Yu. 2018. "Vitamin D deficiency and high serum IL-6 concentration as risk factors for tubal factor infertility in Chinese women". *Nutrition,* 49:24 - 31. doi:10.101 6/j.nut.2017.11.016.

Christesen, H. T., T. Falkenberg, R. F. Lamont and J. S. Jorgensen. 2012. "The impact of vitamin D on pregnancy: a systematic review". *Acta Obstet. Gynecol. Scand.,* 91 (12):1357 - 67. doi: 10.1111/aogs.12000.

Chu, J., I. Gallos, A. Tobias, B. Tan, A. Eapen and A. Coomarasamy. 2018. "Vitamin D and assisted reproductive treatment outcome: a systematic review and meta-analysis". *Hum. Reprod.,* 33 (1):65 - 80. doi: 10.1093/ humrep/dex326.

Ciavattini, A., M. Serri, G. Delli Carpini, S. Morini and N. Clemente. 2017. "Ovarian endometriosis and vitamin D serum levels". *Gynecol. Endocrinol.,* 33 (2):164 - 167. doi: 10.1080/09513590.2016.1239254.

Dastorani, M., E. Aghadavod, N. Mirhosseini, F. Foroozanfard, S. Zadeh Modarres, M. Amiri Siavashani and Z. Asemi. 2018. "The effects of vitamin D supplementation on metabolic profiles and gene expression of insulin and lipid metabolism in infertile polycystic ovary syndrome candidates for in vitro fertilization". *Reprod. Biol. Endocrinol.,* 16 (1): 94. doi: 10.1186/s12958-018-0413-3.

Djurovic, J., G. Stamenkovic, J. Todorovic, N. Aleksic and O. Stojkovic. 2018. "Polymorphisms and haplotypes in VDR gene are associated with female idiopathic infertility". *Hum. Fertil.,* (Camb):1 - 10. doi: 10.108 0/14647273.2018.1515503.

Evans, K. N., L. Nguyen, J. Chan, B. A. Innes, J. N. Bulmer, M. D. Kilby and M. Hewison. 2006. "Effects of 25-hydroxyvitamin D3 and 1,25-dihydroxyvitamin D3 on cytokine production by human decidual cells". *Biol. Reprod.,* 75 (6):816 - 22. doi: 10.1095/ biolreprod.106.054056.

Fatemi, F., A. Mohammadzadeh, M. R. Sadeghi, M. M. Akhondi, S. Mohammadmoradi, K. Kamali, N. Lackpour, S. Jouhari, S. Zafadoust, S. Mokhtar and L. Giahi. 2017. "Role of vitamin E and D3 supplementation in Intra-Cytoplasmic Sperm Injection outcomes of women with polycystic ovarian syndrome: A double blinded randomized placebo-controlled trial". *Clin. Nutr. ESPEN,* 18:23 - 30. doi: 10.1016/j.clnesp.2017.01.002.

Flood-Nichols, S. K., D. Tinnemore, R. R. Huang, P. G. Napolitano and D. L. Ippolito. 2015. "Vitamin D deficiency in early pregnancy". *PLoS One,* 10 (4):e0123763. doi: 10.1371/journal.pone.0123763.

Franasiak, J. M., E. E. Lara and A. Pellicer. 2017. "Vitamin D in human reproduction". *Curr. Opin. Obstet. Gynecol.,* 29 (4):189 - 194. doi:10.1097/gco.0000000000000375.

Gernand, A. D., H. N. Simhan, S. Caritis and L. M. Bodnar. 2014. "Maternal vitamin D status and small-for-gestational-age offspring in women at high risk for preeclampsia". *Obstet. Gynecol.,* 123 (1):40 - 8. doi:10.1097/aog.0000000000000049.

Heaney, R. P. 2008. "Vitamin D in health and disease". *Clin. J. Am. Soc. Nephrol.,* 3 (5):1535 - 41. doi: 10.2215/cjn.01160308.

Holick, M. F. 2007. "Vitamin D deficiency". *N Engl. J. Med.,* 357 (3):266 - 81. doi: 10.1056/NEJMra070553.

Iliuta, F., J. I. Pijoan and R. Matorras. 2018. "Is the association of replete status in vitamin D with better results in IVF demonstrated?" *Hum. Reprod.,* 33 (9):1797. doi: 10.1093/humrep/dey251.

Jiang, L., L. Ji, J. Song and K. Qian. 2019. "The effect of serum vitamin D levels in couples on embryo development and clinical outcomes". *Reprod. Biomed. Online,* 38 (5):699 - 710. doi: 10.1016/j.rbmo.2018.12.036.

Lerchbaum, E. and B. Obermayer-Pietsch. 2012. "Vitamin D and fertility: a systematic review". *Eur. J. Endocrinol.,* 166 (5):765 - 78. doi: 10.1530/eje-11-0984.

Merewood, A., S. D. Mehta, T. C. Chen, H. Bauchner and M. F. Holick. 2009. "Association between vitamin D deficiency and primary cesarean section". *J. Clin. Endocrinol. Metab.,* 94 (3):940 - 5. doi: 10.1210/jc.2008-1217.

Miyashita, M., K. Koga, G. Izumi, F. Sue, T. Makabe, A. Taguchi, M. Nagai, Y. Urata, M. Takamura, M. Harada, T. Hirata, Y. Hirota, O. Wada-Hiraike, T. Fujii and Y. Osuga. 2016. "Effects of 1,25-Dihydroxy

Vitamin D3 on Endometriosis". *J. Clin. Endocrinol. Metab.,* 101 (6):23 71 - 9. doi: 10.1210/jc.2016-1515.

Parikh, G., M. Varadinova, P. Suwandhi, T. Araki, Z. Rosenwaks, L. Poretsky and D. Seto-Young. 2010. "Vitamin D regulates steroidogenesis and insulin-like growth factor binding protein-1 (IGFBP-1) production in human ovarian cells". *Horm. Metab. Res.,* 42 (10):754 - 7. doi: 10.1055/s-0030-1262837.

Polyzos, N. P., E. Anckaert, L. Guzman, J. Schiettecatte, L. Van Landuyt, M. Camus, J. Smitz and H. Tournaye. 2014. "Vitamin D deficiency and pregnancy rates in women undergoing single embryo, blastocyst stage, transfer (SET) for IVF/ICSI". *Hum. Reprod.,* 29 (9):2032 - 40. doi: 10.1 093/humrep/deu156.

Shand, A. W., N. Nassar, P. Von Dadelszen, S. M. Innis and T. J. Green. 2010. "Maternal vitamin D status in pregnancy and adverse pregnancy outcomes in a group at high risk for pre-eclampsia". *Bjog,* 117 (13):1593 - 8. doi: 10.1111/j.1471-0528.2010.02742.x.

Soheilykhah, S., M. Mojibian, M. Rashidi, S. Rahimi-Saghand and F. Jafari. 2010. "Maternal vitamin D status in gestational diabetes mellitus". *Nutr. Clin. Pract.,* 25 (5):524 - 7. doi: 10.1177/ 0884533610379851.

Thomson, R. L., S. Spedding and J. D. Buckley. 2012. "Vitamin D in the aetiology and management of polycystic ovary syndrome". *Clin. Endocrinol.,* (Oxf) 77 (3):343 - 50. doi: 10.1111/j.1365-2265.2012.044 34.x.

Vigano, P., D. Lattuada, S. Mangioni, L. Ermellino, M. Vignali, E. Caporizzo, P. Panina-Bordignon, M. Besozzi and A. M. Di Blasio. 2006. "Cycling and early pregnant endometrium as a site of regulated expression of the vitamin D system". *J. Mol. Endocrinol.,* 36 (3):415 - 24. doi: 10.1677/jme.1.01946.

Yoshizawa, T., Y. Handa, Y. Uematsu, S. Takeda, K. Sekine, Y. Yoshihara, T. Kawakami, K. Arioka, H. Sato, Y. Uchiyama, S. Masushige, A. Fukamizu, T. Matsumoto and S. Kato. 1997. "Mice lacking the vitamin D receptor exhibit impaired bone formation, uterine hypoplasia and growth retardation after weaning". *Nat. Genet.,* 16 (4):391 - 6. doi: 10.1038/ng0897-391.

Zarnani, A. H., M. Shahbazi, A. Salek-Moghaddam, M. Zareie, M. Tavakoli, J. Ghasemi, S. Rezania, A. Moravej, E. Torkabadi, H. Rabbani and M. Jeddi-Tehrani. 2010. "Vitamin D3 receptor is expressed in the endometrium of cycling mice throughout the estrous cycle". *Fertil. Steril.,* 93 (8):2738 - 43. doi: 10.1016/ j.fertnstert.2009.09.045.

Zhao, J., S. Liu, Y. Wang, P. Wang, D. Qu, M. Liu, W. Ma and Y. Li. 2019. "Vitamin D improves in-vitro fertilization outcomes in infertile women with polycystic ovary syndrome and insulin resistance". *Minerva Med.,* 110 (3):199 - 208. doi: 10.23736/s0026-4806.18.05946-3.

Zhao, Jing, Xi Huang, Bin Xu, Yi Yan, Qiong Zhang and Yanping Li. 2018. "Whether vitamin D was associated with clinical outcome after IVF/ICSI: a systematic review and meta-analysis". *Reproductive biology and endocrinology : RB&E,* 16 (1):13 - 13. doi: 10.1186/ s12958 -018-0324-3.

In: Vitamin D as Progesterone-Like Hormone ISBN: 978-1-53617-516-5
Editor: Vittorio Unfer

Chapter 7

A Support for the Luteal Phase: Supplementation with Vitamin D, Inositol and Melatonin

Lidia Di Cerbo[1,*], MD, Conisha Holloman[2], MD and Monica Longo[2], MD, PhD

[1]Mother Infant Department,
University of Modena and Reggio Emilia, Modena, Italy
[2]McGovern Medical School University of Texas Health Science Center at Houston (UTHealth), Maternal-Fetal Medicine, Houston, Texas, US

Abstract

The synthesis and secretion of progesterone (P4) by the corpus luteum in the luteal phase are of crucial importance in determining the implantation of the blastocyst and the maintenance of pregnancy in the early first trimester. Luteal phase deficiency (LPD) is associated with many fertility disorders and higher pregnancy losses. Although several studies, in human and animal models, have been performed to understand luteal biology, there is still a lack of knowledge on luteal phase defects and

[*] Corresponding Author's Email: lidiadicerboldc@gmail.com.

controversies regarding the diagnosis and potential treatment of luteal inadequacy. Vitamin and non-vitamin supplements such as vitamin D, inositol and melatonin are newer innovative treatments that could provide support for the luteal phase defects.

Keywords: luteal phase, vitamin D, inositol, melatonin, supplementation

INTRODUCTION

The luteal phase is characterized by the presence of the corpus luteum (CL), a transient endocrine gland that develops after the rupture of the mature Graafian follicle to release ova. The CL synthesizes and secretes progesterone (P4), which is necessary for the implantation of the blastocyst and the maintenance of pregnancy in the early first trimester until the placenta begins to function. Removal of the CL before the development of adequate placental function results in spontaneous pregnancy loss. Given the importance of ovarian P4 production for implantation and early pregnancy support, the potential for ovarian inadequacy as a cause of infertility or pregnancy failure seems plausible. Studies supporting the need for adequate P4 during the luteal phase (Baird et al., 1997) suggest that cycles resulting in conception have a more rapid rise of P4 and higher mid-luteal estrogen and P4 levels (Baird et al., 1991). However, similar increased in luteal phase P4 levels have been observed in cycles with normal and biochemical pregnancies, suggesting that pregnancy losses may result from other causes other than ovarian hormonal deficiency (Baird et al., 1991).

Luteal phase deficiency (LPD) is a condition first described in 1949 (Jones, 1949). It is characterized by insufficient endogenous P4 levels needed to maintain a functional secretory endometrium and allow normal embryo implantation and growth. Controversy regarding the clinical significance of LPD is due to the lack of a reliable test to diagnose the disorder.

In previous years, LPD was associated with infertility (Moszkowski et al., 1962; Blacker et al., 1997); first-trimester pregnancy loss (Swyer and Daley, 1952); short cycle (Strott et al., 1970; Suh and Betz, 1993), premenstrual spotting (Muechler et al., 1987); anorexia, starvation and eating disorders (Pirke et al., 1985); excessive exercise (Cree, 1998); stress (Kajantie and Phillips, 2006); obesity and polycystic ovary syndrome (PCOS) (Filicori et al., 1991); endometriosis (Cunha-Filho et al., 2003); aging (Prior, 2005); ovulation stimulation alone (Daly et al., 1983; Olson et al., 1983); ovulation induction with or without gonadotropin-releasing agonists and assisted reproductive technology (ART) (Tavaniotou et al., 2002). Although the association of LPD with infertility have not been proven, it is essential to understand luteal biology for better infertility management (Pfeifer et al., 2015; Alliende et al., 2018).

The premature onset of menses was recognized as an indication of a luteal-phase deficiency of P4 production, which was corrected by exogenous P4 administration (Jones, 1949). The luteal phase in a natural cycle is considered abnormal when serum mid-luteal P4 levels decrease to less than 10 ng/ml, but later it was shown to have no correlation between plasma levels of P4, tissue levels and the histological pattern (Jordan et al., 1994). The most reasonable consensus on a luteal phase defect was suggested as a lag of more than 2 days in endometrial histological development compared to the expected day of the cycle, based on the criteria of Noyes (Bourgain et al., 1994). In the early 1980s, two famous pioneers in *in-vitro* fertilization (IVF), Robert G. Edwards and Howard W. Jones, reported a shortened luteal phase when human menopausal gonadotropins (hMG) alone were used to induce multiple follicular growth for *in-vivo* and *in-vitro* fertilization. The use of gonadotropin-releasing hormone agonists for a combined therapy with hMG suppressed the output of preovulatory luteinizing hormone (LH) and was deleterious for embryo quality. Supraphysiological levels of steroids such as estradiol, secreted by a high number of corpora lutea during the early luteal phase of stimulated cycles, inhibited the LH release via negative feedback actions at the level of the hypothalamic–pituitary axis and induced premature luteolysis. LH support during the early luteal phase is entirely responsible for the maintenance of the steroidogenic activity necessary to prepare the endometrium for embryo implantation. Thus, low implantation and low pregnancy rates were observed when adequate exogenous luteal-phase support (LPS) was not provided (Fatemi et al.,

2007). P4 was recognized as the best method to provide LPS in ART, as was associated with lower rates of ovarian hyperstimulation syndrome, more prolonged pregnancies and higher live birth rates compared to placebo (van der Linden et al., 2015).

Primate research involved in understanding the CL is challenging. Although there are similarities between CL of women and non-human primates as well other species, such as rodents, there are still many species-specific differences in luteal biology. Therefore, research into the human CL suffers from limitations of tissue availability and access and appropriate animal models. In the case of LPD, there is a lack in the ability to differentiate luteal tissues and luteal phase that are likely to regress prematurely or secrete P4 insufficiently compared to those that are normal (Parisi et al., 2014).

One solution to address LPD may be found in the luteal tissue genomics and proteomics research. Indeed, there are many unanswered questions, which are still being explored.

The initial step to address these questions will be to develop non-human primate models to mimic the human luteal phase defects. Earlier studies have documented the usefulness of non-human primate for investigation of LPD (Duncan et al., 2009) and generated the following data:

1. Observational approach that compares the expression of genes and proteins between a functional CL to the regressing CL in the late luteal phase.
2. Manipulation of the master regulator of CL such as the LH/hCG prior to the CL removal, either by inducing luteolysis through administration of GnRH receptor antagonists or rescue of CL from luteolysis by readministration of LH mimicking the hormonal change of early pregnancy.
3. *In vitro* culture of luteal cells by reductionist approach.

Luteal tissue is complex with many distinct cell populations. The ideal way to arrive at solid conclusions is to dissect complex tissue into simpler components, interrogating each separately. Using currently available primate luteal gene expression arrays and bioinformatics will be the first step in a large-scale effort to analyze a variety of luteal phase defects. Behind these gene expression analyses stands a generation of proteomic science, in which in a LPD, analysis of the spectrum of proteins in a CL during different functional states, from induced and spontaneous luteolysis, will provide critical information about functional genomics and proteomics formation.

In light of the above findings, the lack of clear information about the luteal phase defects and the need of more studies, there are several controversies regarding the diagnosis and potential treatment of luteal inadequacy. Maternal dietary habits during the preconception period are considered increasingly important to cover the nutritional needs of the fetus and placenta during pregnancy (Parisi et al., 2014), and to optimize maternal and infant health (Huijgen et al., 2014). Consequently, several medical societies have published preconception care guidelines to manage a woman's dietary intake and nutritional status before conception. Despite these guidelines, recent studies showed that preconception dietary intake is still suboptimal (Blumfield et al., 2013). We have focused on the possibility of providing support for the luteal phase by administering vitamin and non-vitamin supplements: vitamin D, inositol and melatonin.

VITAMIN D

Vitamin D is a prohormone produced photochemically in the skin from 7-dehydrocholesterol. The molecular structure of vitamin D is closely allied to that of classic steroid hormones (i.e., estradiol, cortisol, and aldosterone) due to the same root cyclopentano-perhydro-phenanthrene ring structure. Technically, vitamin D is a secosteroid because one of its structural rings has a broken carbon-carbon bond.

Vitamin D is known to be involved in calcium-phosphate homeostasis and bone metabolism (Yoshizawa et al., 1997; Panda et al., 2001). Emerging data identify critical roles for vitamin D in a variety of other biological processes including regulation of cellular growth and differentiation and metabolic modulations specifically involving insulin action (Heaney, 2008).

Vitamin D has many beneficial roles in a spectrum of pathologic processes, including autoimmunity, insulin resistance, cardiovascular disease, and malignancies.

Biological actions of vitamin D are mediated through the vitamin D receptor (VDR) (Johnson and DeLuca, 2001). This receptor is a member of the steroid/thyroid nuclear hormone receptor superfamily and has been demonstrated in calcium-regulating tissues, intestines, skeleton, parathyroid glands, and reproductive tissues such as ovary, uterus, placenta, testis, and the hypophysis (Lerchbaum and Obermayer-Pietsch, 2012; Luk et al., 2012). For those reasons, among many physiologic processes influenced by vitamin D, critical roles have been found in reproductive physiology (Hurley and Doane, 1989; Harkness and Bonny, 2005; Johnson and DeLuca, 2002). For example, a diet lacking in vitamin D has been found to reduce fertility by 45%-75% in rats. Experiments investigating the significance of vitamin D for fertility and reproductive capacity have demonstrated a non-critical role for successful female reproduction but did find compromised mating behavior, reduced fertility rates, decreased litter sizes, and impaired neonatal growth in vitamin D–deficient female rats (Halloran and DeLuca, 1980). Knock-out mice that cannot convert vitamin D to its active form show estrus cycle disturbances including arrest of follicular development, prolonged estrous cycles, and anovulation (Dicken et al., 2012).

Vitamin D deficiency is also a risk factor for PCOS. Among women with PCOS, the supplementation with vitamin D can normalize menstrual cycles and improve ovarian folliculogenesis (Irani and Merhi, 2014; Rashidi et al., 2009). A recent review (Menichini and Facchinetti, 2019) presented the effects of vitamin D supplementation on glucose metabolism, insulin resistance, hyperlipidemia and hormonal status in women with PCOS. Although the results are quite heterogeneous, the better outcomes have been obtained with high (therapeutic) doses of vitamin D (≥4000 IU/d) administered for a period of at least 12 weeks. It showed that vitamin D supplementation helped restoring physiological serum 25-hydroxyvitamin D (25(OH)D) levels in women with vitamin D deficiency. Indeed, taking into account that approximately 67–85% of women with PCOS have insufficient levels of vitamin D (He et al., 2015), supplementation should be recommended to all women with PCOS. However, there are only very few studies of vitamin D and ovarian function in healthy women. Other studies found an association between 25(OH)D, the accepted biomarker of vitamin D status, and long menstrual cycles (Jukic et al., 2016) or irregular cycles (Jukic et al., 2015). Long or irregular menstrual cycles can require hormonal therapy and medical intervention, with their associated costs. Long menstrual cycles (>35 days) are more likely to be anovulatory (Mumford et al., 2012). If vitamin D is associated with menstrual cycle length, it would suggest that the human reproductive axis could be responsive to vitamin D supplementation.

Both cycling and early pregnant endometrium represents an extra-renal site of vitamin D synthesis; moreover, the effect of vitamin D at the uterine level is thought to be via the VDR through either the regulation of target genes or the hormonal effects on the local immune response (Viganò et al., 2006). Furthermore, an adequate preconception of vitamin D status is important as vitamin D deficiency in the mother has been linked to various issues in the pregnancy such as preeclampsia (Bodnar et al., 2007) and gestational diabetes (Zhang et al., 2008) and in their offspring in term of low birth weight (Morley et al., 2009), poor bone growth (Javaid et al., 2006), and asthma (Camargo et al., 2007; Devereux et al., 2007). Sunlight is the predominant source of vitamin D (Brouwer-Brolsma et al., 2016), whereas dietary vitamin D can only be obtained through the consumption of a limited number of foods, such as fatty fish, fat spreads, oils, liver, meat, eggs, and dairy products. For pregnant and lactating women, optimal serum 25(OH)D levels have not yet been defined. A 2009 review (Mulligan et al., 2009) has recommended that women at risk of vitamin D deficiency should be monitored and if deficient start a daily intake of 1000 IU/day. However, another study found that women who are vitamin D deficient at the beginning of pregnancy may remain deficient even with daily doses of 1000 IU/day (Datta et al., 2002). The Canadian Paediatric Society in 2007 has recommended a daily dose of 2000 IU in pregnant and lactating women (CPS, 2007). However, doses over 2000 IU may be required to maintain serum 25(OH)D levels over 80 nmol/L (Hollis and Wagner, 2006). The major circulating form of vitamin D is 25(OH)D, which have been used as a measure of adequacy for vitamin D, as they reflect intake from the diet coupled with the amount contributed by cutaneous synthesis. The reference range of the total 25(OH)D level is 20-100 ng/mL. In 2010, the Institute of Medicine (IOM) recommended 600 IU/day vitamin D supplementation between 1 to 70 years of age (Ross et al., 2011). Recommendations from the Institute of Medicine suggest that an average daily intake of 400–800 IU, or 10–20 micrograms, is adequate for 97.5% of individuals.

This dosage is only adequate for people who have normal ranges of 25(OH)D (Bischoff-Ferrari, 2009). Unfortunately, the exact amounts of vitamin D supplementation for populations at high-risk for deficiency in the preconception period is still unclear. Because of the influential role of Vitamin D in reproductive women and their future pregnancies, the time and minimum dosage required to reach normal vitamin D ranges is necessary. However, the evidence currently available to assess the benefits and harms of the use of vitamin D supplementation alone in pregnancy for improving maternal and infant health outcomes is limited.

INOSITOL

The widespread use of inositol compounds in the clinical gynecologic practice started in 1850, when Johannes Joseph Scherer isolated from the muscle a hexa-hydroxycyclohexane compound that he named Inositol. Inositol is a carbohydrate and its structure allows the formation of 9 different stereoisomers, 2 of them are predominantly found in eukaryotic cells: myo-inositol (MI) and D-chiroinositol (DCI). These two stereoisomers are the main ones present in humans with MI being the most abundant form in living cells and the most distributed in biological systems. Their plasma ratio is MI/DCI 40:1 (Unfer and Porcaro, 2014).

Inositol pyrophosphates (PP-InsPs) exhibit activity as a protein pyrophosphorylation and has strong electrostatic interactions with target proteins. Many of Inositol activities derive from its ability to transfer information into the cell and regulate the metabolic signaling acting as second messengers.

MI and its derivatives are a unique group of molecules, which can be found in various foods. In plants, it is generally represented in the form of hexaphosphate and phytic acid, or its salts (phytates). The greatest amount of MI is present in fresh fruits, vegetables, and foods containing seeds (beans, grains and nuts). High phytic acid contents is mostly found in almonds, walnuts, Brazil nuts and oats; bran contain more MI than cereals derived from any other grains. Among vegetables, the highest contents are observed in the beans and peas, with leafy vegetables being the poorest vegetable sources of MI. Among fruits, cantaloupe and citrus fruits (with the exception of lemons) have extraordinarily high contents of MI. It was once considered a member of the vitamin B complex but cannot be deemed as a 'true' essential nutrient since it can be synthesized by the human body from glucose (mostly in liver and kidney) (Clements and Darnell, 1980).

In a two-step reaction, glucose-6-phosphate is first isomerized by inositol-3-phosphate synthase 1 into inositol-3-phosphate, which is then dephosphorylated by the inositol-monophosphatase into MI (Ju et al., 2004). There is limited effectiveness of this reaction, and thus the major part of inositol comes from exogenous food sources. Both MI and DCI are vital, biologically active molecules, in the cell. They are components of cell membrane phospholipids, with both structural and functional roles (Di Paolo and De Camilli, 2006). Cell membrane phospholipids are the source of inositol triphosphate, diacylglycerol and inositol phosphoglycans that act as second messengers of many metabolic pathways. Some of these pathways include those depending on thyroid stimulating hormone (TSH), follicular stimulating hormone (FSH), luteinizing hormone (LH) and insulin. The only difference between MI and DCI is a hydroxyl group, but their biological properties are significantly different. Both MI and DCI are regarded as insulin sensitizers' compounds, but act through different mechanisms. MI regulates the cellular uptake of glucose, while DCI mainly stimulates glycogen synthase (GS) and regulates steroidogenesis. In glycogen-storage tissues (liver, muscle, fat) the DCI to MI ratio is higher than in tissues of high glucose utilization (heart, ovary) (Larner et al., 2010). Conversion of MI to DCI is regulated by an insulin-dependent epimerase. The impairment of the epimerase function may lead to inositol imbalance and is associated with conditions such as type 2 diabetes and PCOS (Sun et al., 2002). Receptor bound glycoprotein hormones stimulate two pathways: the primary, the cyclic adenosine 3′,5′-monophosphate (cAMP) dependent, activated by the Gs protein, and the latter, activated by the Gq protein, is dependent on the inositol derivatives (Gloaguen et al., 2011). The inositol pathway activation results in the efflux of calcium ions from the smooth endoplasmic reticulum into the cytosol. When higher concentrations of the hormone and receptor density are present, the Gq protein is activated (Allen et al., 2011). This dual signaling has an impact on ovulation (Ewa et al., 2016) and human reproduction (Papaleo et al., 2009). FSH receptors stimulation, in the early follicular phase, activates the cAMP pathway. A high concentration of cAMP maintains oocyte in prophase 1 and stimulates proliferation and growth of granulosa cells. The ovulatory LH surge involves the activation of the inositol pathway. The change of cAMP promotes the resumption of meiosis and releases the mature oocyte. Therefore, despite the similar role of MI and DCI on insulin resistance (Genazzani et al., 2008; Giordano et al., 2011), it has been shown that an elevated concentration of MI in follicular fluid (FF) appears to play a specific role in follicular maturation and provides a marker of good oocyte quality (Chiu et al. 2002). Both MI and DCI are insulin mimetic when administered *in vivo* enhancing the physiological insulin-receptor activity and reducing glucose levels in serum. The administration of MI and DCI has been related to an improvement of ovulation in most PCOS women (Cheang et al., 2008; Costantino et al. 2009; Marcin et al., 2019). Indeed, MI supplementation reduces serum insulin levels, decreases serum testosterone, and restores ovulation (Nestler, 1997). Moreover, experiments on mice oocytes have shown that exposure to MI positively affects the meiotic progression of germinal vesicle by enhancing the intracellular $Ca2^{+}$ oscillation (Papaleo et al., 2009; Chiu et al., 2003). On the other hand, it is well known that DCI glycans administration is related to the activation of glycogen synthase (Larner, 2001). The MI/DCI physiological ratio is specific in each tissue and the enzymatic conversion of the two isomers is affected by the condition of insulin resistance (Sun et al., 2002). The 40:1 MI/DCI ratio is the best for PCOS therapy aimed at restoring ovulation and normalizing important parameters in these patients. The other formulations are less effective. In particular, a decreased activity was observed when the 40:1 ratio was modified in favour of DCI (Nordio et al., 2019). Although inositol dietary deficiency in the general population has not been demonstrated at the serum level, several findings are emerging regarding the impact of inositol supplementation in periconceptional period and in early phases of pregnancy. Inositol supplementation may have an improving effect on metabolic reassessment of early and late pregnancy (Wilson et al., 2009; Khandelwal et al., 1998), and a potentially therapeutic role in human reproduction. In the myriad of ''natural measures'' focused at optimizing periconceptional maternal health status and metabolic balance, inositol dietary supplementation could not be excluded from the discussion. Evidence regarding the beneficial effects of this molecule in improving glucidic and lipidic profiles, fertility, and the success rate of ART in women affected by PCOS are strong and well confirmed. MI and DCI have both shown a possible role in restoring ovulation in PCOS patients. The combination of MI and DCI in a physiological ratio seems more effective rather than single compound alone in restoring ovulation and ameliorating metabolic factors (Unfer et al., 2016). Focusing on gestational diabetes and its impact on pregnancy (considering both the mother and the fetus) and peripartum outcomes (IADPSG, 2010; D'Anna et al., 2015) the validation of a ''non pharmacological'' treatment as a preventive and therapeutic agent in the absence of side effects may implicate inositol as a largely beneficial molecule.

MELATONIN

Melatonin is an indoleamine (N-acetyl-5-methoxytryptamine) involved in the regulation of circadian rhythms, sleep and food-intake (Arendt, 2006; Lieberman et al., 1984; Mustonen et al., 2001). This "hormone of darkness" is primarily synthesized in the pineal gland during obscurity and its release is inhibited by light. In particular, it peaks between 2 and 4 a.m. and gradually falls during the second half of the night.

The initial precursor of melatonin biosynthesis is the amino acid tryptophan. Pinealocytes take up tryptophan from the blood and convert it to serotonin through hydroxylation and decarboxylation; serotonin is then converted to N-acetyl-serotonin by the enzyme arylalkylamine N-acetyltransferase (NAT). N-acetyl-serotonin is subsequently methylated to form melatonin; this step requires the enzyme hydroxyindole-O-methyltransferase (HIOMT). The pineal activity of NAT usually exhibits the same circadian rhythm as blood and pineal melatonin levels are being suppressed by light. In vertebrates, apart from the pineal gland, many peripheral tissues express both melatonin and its receptors (MT1 or MTNR1A and MT2 or MTNR1B) such as brain, retina, the cardiovascular system, liver, gallbladder, gastrointestinal tract, kidney, immune cells, adipocytes, prostate and breast epithelial cells, myometrium, skin and ovary. However, its rhythmic production occurs only in the pineal gland and retina (Niles et al., 1999). Melatonin is detectable in virtually every bodily compartment including human preovulatory FF where its concentration is significantly higher than that in the peripheral serum (Nakamura et al., 2003). Although it has been assumed that melatonin measured in human preovulatory FF is derived from the general circulation (Brzezinsk et al., 1987), it may also be synthesized in the ovary, as both NAT and HIOMT are present in ovarian tissue (Itoh et al., 1999). However, the current opinion is that the bulk of the melatonin detected in the ovary and preovulatory FF is derived from the circulation. One study demonstrated that melatonin administration (3 mg tablet orally) to infertile women resulted in high melatonin concentrations in their FF (after melatonin treatment: 432 ± 260 pg/mL vs. control: 112 ± 51 pg/mL) (Tamura et al., 2008). In recent years, several studies proposed a link between melatonin and reproduction, with its involvement in ovarian physiology, including steroidogenesis (Tamura et al., 2009). Indeed, serum levels of this hormone have been found higher in the luteal phase than the follicular phase of the menstrual cycle (Wetterberg et al., 1976) and recently demonstrated the ability of melatonin to restore a normal menstrual cycle (Tagliaferri et al., 2018). In addition, several studies reported a correlation between this hormone and sex steroid production. In particular, melatonin was able to increase P4 production by growing and luteinized granulosa luteal cells (GCs) (Woo et al., 2001; Webley and Leidenberger, 1986). On the contrary, some authors showed no or negative effects on P4 release (Bódis et al., 2001; Murayama et al., 1997). *In vitro* models show that melatonin regulates steroidogenesis, folliculogenesis, and oocyte maturation within the ovary (Tanavde and Maitra, 2003). Melatonin also seems to protect follicles against oxidative stress, due to its documented properties as an antioxidant and free radical scavenger. Melatonin also rescues follicles from atresia by promoting the correct follicular maturation and ovulation (Jain et al., 2013). In addition, melatonin stimulates a number of enzymes that are involved either in metabolizing potentially reactive oxygen species to harmless molecules or in inducing the synthesis of endogenous antioxidants (Barlow-Walden et al., 1995) products. Several studies in rats also demonstrated a role of melatonin in the regulation of gonadotropin-releasing hormone (GnRH) secretion in the hypothalamus and in the pituitary gland (Tamura et al., 2013). Interestingly, the reduction in melatonin levels with pinealectomy seems to induce the development of some characteristics of PCOS in rats. Conversely, studies in rat models of PCOS have shown increased P4 and androgen production in preantral follicles after incubation with melatonin for 12 days (Prata et al., 2004). However, the role of melatonin in influencing the hypothalamic–pituitary–gonadal axis in humans is still unclear.

There are reports of increased melatonin levels in the luteal phase compared with the follicular phase of the menstrual cycle (Wetterberg et al., 1976; Brun et al., 1987). In humans, melatonin binding sites have been detected in granulosa cells–luteal cells, and melatonin directly stimulates the secretion of P4 by human GCs or luteal cells (Hsu et al., 1996). It may act at the level of the ovary to modify its luteal function. In addition to melatonin's stimulation of P4 production by GCs–luteal cells, it also remarkably increases mRNA expression of the LH (but not FSH) receptor in human GCs/luteal cells, while inhibiting expression of GnRH and its receptor (Woo et al., 2001). Melatonin enhances hCG-stimulated P4 secretion from these cells, possibly by the increased expression of LH receptors.

As reviewed, evidence suggests the involvement of melatonin in ovarian physiology: follicular development, ovulation, oocyte maturation, and luteal function. Moreover, a melatonin deficiency seems to be involved in the pathophysiology of PCOS, endometriosis, and premature ovarian failure (POF). The clinical use of melatonin in humans has been explored and the findings highlight new opportunities for the management of several diseases including cancer, immunological disorders, Alzheimer's disease, diabetes, and viral infections (Pappolla et al., 2000). In humans, the safety of exogenous melatonin has been revealed in many studies. The usage of melatonin for ovarian diseases, such as PCOS, alone (Mokhtari et al., 2019) or in combination with MI, have shown to enhance, synergistically, oocyte and embryo quality (Pacchiarotti et al., 2015). Some other clinical trials on melatonin supplementation proved its therapeutic benefit, in particular melatonin ameliorates the activity of MI and folic acid by improving oocyte quality and pregnancy outcome in women with low oocyte quality history (Rizzo et al., 2010). In another recent trial, treatment with myo-inositol and melatonin, in infertile women with poor oocyte quality, improved ovarian stimulation protocols and pregnancy outcomes (Unfer et al., 2011).

CONCLUSION

In this chapter we extensively reviewed the data to understand how these intrinsically produced molecules, such as vitamin D, inositols and melatonin are efficient in helping the LPD. In particular, several clinical trials have shown that supplementation with MI and DCI, alone or in association with melatonin, as well as vitamin D supplementation, is a newer approach to improve oocyte quality and IVF outcomes in both PCOS patients and normal subjects. Due to the lack of data we are not able to specify the exact daily dose. In general, vitamin D supplementation with 10-15 μg/day is recommended for women in the preconception period or already pregnant, particularly during the winter months. Inositol supplementation, namely in the recommended dose of 4gr/day, is helpful to improve insulin resistance and fertility in women with PCOS. For melatonin, the recommended dose to use (alone or in combination with inositols) is 1 mg/ day, preferably at night-time prior to sleep.

References

Allen, M. D., S. Neumann, M. C. Gershengorn. 2011. "Occupancy of both sites on the thyrotropin (TSH) receptor dimer is necessary for phosphoinositide signaling." *FASEB J;* 25:3687–94.

Arendt, J. 2006. "Melatonin and human rhythms." *Chronobiol Int* 23(1–2):21–37 (Review).

Baird, D. D., A. J. Wilcox, C. R. Weinberg, et al. 1997. "Preimplantation hormonal differences between the conception and non-conception menstrual cycles of 32 normal women." *Hum Reprod,* 12, pp. 2607-2613.

Baird, D. D., C. R. Weinberg, A. J. Wilcox, et al. 1991. "Hormonal profiles of natural conception cycles ending in early, unrecognized pregnancy loss." *J Clin Endocrinol Metab,* 72, pp. 793-800.

Barlow-Walden, L. R, R. J. Reiter, M. Abe, M. Pablos, A. Menendez-Pelaez, L. D. Chen, et al. 1995. "Melatonin stimulates brain glutathione peroxidase activity." *Neurochem Int;* 26:497–502.

Bischoff-Ferrari, H. 2009. "Vitamin D: what is an adequate vitamin D level and how much supplementation is necessary?" *Best Pract Res Clin Rheumatol;* 23(6):789–95.

Blacker, C. M., K. A. Ginsburg, R. E. Leach, J. Randolph, K. S. Moghissi. 1997. "Unexplained infertility: evaluation of the luteal phase; results of the National Center for Infertility Research at Michigan." *Fertil Steril,* 67, pp. 437-442.

Blumfield, M. L., A. J. Hure, L. Macdonald-Wicks, R. Smith, C. E. Collins. 2013. "A systematic review and meta-analysis of micronutrient intakes during pregnancy in developed countries." *Nutr. Rev.* 71, 118–132.

Bódis, J., M. Koppán, L. Kornya, H. R. Tinneberg, A. Török. 2001. "Influence of melatonin on basal and gonadotropin-stimulated progesterone and estradiol secretion of cultured human granulosa cells and in the superfused granulosa cell system." *Gynecol Obstet Invest* 52(3):198–202.

Bodnar, L. M., J. M. Catov, H. N. Simhan, M. F. Holick, R. W. Powers, J. M. Roberts. 2007. "Maternal vitamin D deficiency increases the risk of preeclampsia." *J. Clin. Endocrinol. Metab,* 92, 3517–3522.

Bourgain, C, J. Smitz, M. Camus, et al. 1994. "Human endometrial maturation is markedly improved after luteal supplementation of gonadotrophin-releasing hormone analogue/human menopausal gonadotrophin stimulated cycles." *Hum Reprod;* 9:32–40.

Brouwer-Brolsma, E. M., A. M. Vaes, N. L. van der Zwaluw, J. P vanWijngaarden, K. M. Swart, et al. 2016. "Relative importance of summer sun exposure, vitamin D intake, and genes to vitamin D status in dutch older adults: The b-proof study." *J. Steroid Biochem. Mol. Biol.* 164, 168–176.

Brun, J., B. Claustrat, M. David. 1987. "Urinary melatonin, LH oestradiol, progesterone excretion during the menstrual cycle or in women taking oral contraceptives." *Acta Endocrinol (Copenh);* 116:145–9.

Brzezinsk, A., M. M. Seibel, H. J. Lynch, M. H. Deng, R. J. Wurtman. 1987. "Melatonin in human preovulatory follicular fluid." *J Clin Endocrinol Metab*; 64:865–7.

Camargo, C. A., S. L. Rifas-Shiman, A. A. Litonjua, J. W. Rich-Edwards, S. T. Weiss, D. R. Gold, K. Kleinman, M. W. Gillman. 2007. "Maternal intake of vitamin D during pregnancy and risk of recurrent wheeze in children at 3 y of age." *Am. J. Clin. Nutr.* 85, 788–795.

Canadian Paediatric Society. 2007. "Vitamin D supplementation: recommendations for Canadian mothers and infants [Position Statement (FNIM 2007-01)]." *Paediatr. Child Health,* 12, 583–589.

Cheang, K., J. Baillargeon, P. Essah, R. Ostlund, T. Apridonize, L. Islam, J. Nestler. 2008. "Insulin-stimulated release of D-chiro-inositol-containing inositolphosphoglycan mediator correlates with insulin sensitivity in women with polycystic ovary syndrome." *Metabolism;* 57: 1390-1397.

Chiu, T. T., M. S. Rogers, C. Briton-Jones, C. Haines. 2003. "Effects of myo-inositol on the in-vitro maturation and subsequent development of mouse oocytes." *Hum Reprod;* 18: 408-416.

Chiu, T. T., M. S. Rogers, E. L. Law, C. M. Briton-Jones, L. P. Cheung, C. J. Haines. 2002. "Fluid and serum concentrations of myo-inositol in patients undergoing IVF: relationship with oocyte quality." *Hum Reprod;* 17: 1591-1596.

Clements, R.S., B. Jr Darnell. 1980. "Myo-inositol content of common foods: development of a high myo-inositol diet." *Am J Clin Nutr* 33:1954–67.

Costantino, D., G. Minozzi, E. Minozzi, C. Guaraldi. 2009. "Metabolic and hormonal effects of myo-inositol in women with polycystic ovary syndrome: a doubleblind trial." *Eur Rev Med Pharmacol Sci;* 13: 105-110.

Cunha-Filho, J. S., J. L. Gross, C. A. Bastos de Souza, N. A. Lemos, C. Giugliani, F. Freitas, & E. P. Passos (2003). Physiopathological aspects of corpus luteum defect in infertile patients with mild/minimal endometriosis. *Journal of assisted reproduction and genetics,* 20(3), 117–121. doi:10.1023/a:1022625106489.

D'Anna, R., A. Santamaria, F. Corrado, A. D. Benedetto, E. Petrella E, Facchinetti F. 2015. [11-OR]. "Myo-inositol in the prevention of gestational diabetes and its complications." *Pregnancy Hypertens.;* 5(1):6.

Daly, D. C., C. A. Walters, C. E. Soto-Albors, D. H. Riddick. 1983. "Endometrial biopsy during treatment of luteal phase deficiencies is predictive of therapeutic outcome." *Fertil Steril,* 40, pp. 305-310.

Datta, S., M. Alfaham, D. P. Davies, F. Dunstan, et al. 2002. "Vitamin D deficiency in pregnant women from a non- European ethnic minority population – an interventional study." *BJOG,* 109, 905–908.

De Cree, C., 1998. "Sex steroid metabolism and menstrual irregularities in the exercising female. A review." *Sports Med,* 25, pp. 369-406.

Devereux, G., A. A. Litonjua, S. W. Turner, L.C. Craig, G. McNeill, et al. 2007. "Maternal vitamin D intake during pregnancy and early childhood wheezing." *Am. J. Clin. Nutr.* 85, 853–859.

Di Paolo, G., P. De Camilli. 2006. "Phosphoinositides in cell regulation and membrane dynamics." *Nature* 443:651–7.

Dicken, C. L., D. D. Israel, J. B. Davis, Y. Sun, J. Shu, J. Hardin, G. Neal-Perry. 2012. "Peripubertal vitamin D(3) deficiency delays puberty and disrupts the estrous cycle in adult female mice." *Biol Reprod;*87(2):51.

Duncan, WC, M. Myers, R. E. Dickinson, S. van den Driesche, H. M. Fraser. 2009. "Paracrine regulation of luteal development and luteolysis in the primate." *Anim Reprod;* 6:34–46.

Ewa, M., A. Czyzyk, B. Meczekalski & A. D. Genazzani. 2016. "Inositol and human reproduction. From cellular metabolism to clinical use." *Gynecological Endocrinology,* doi: 10.1080/09513590.2016.1188282.

Fatemi, H. M., B. Popovic-Todorovic, E. Papanikolaou, P. Donoso, P. Devroey. 2007. "An update of luteal phase support in stimulated IVF cycles." *Hum Reprod Update;* 13:581–90.

Genazzani, A. D., C. Lanzoni, F. Ricchieri, V. Jasonni. 2008. "Myo-inositol administration positively affects hyperinsulinemia and hormonal parameters in overweight patients with polycystic ovary syndrome." *Gynecol Endocrinol;* 24: 139-144.

Giordano, D., F. Corrado, A. Santamaria, S. Quattrone, B. Pintaudi, A. Di Benedetto, R. D'Anna. 2011. "Effects of myo-inositol supplementation in postmenopausal women with metabolic syndrome: a perspective, randomized, placebo-controlled study." *Menopause;* 18: 102-104.

Gloaguen, P., P. Cre´pieux, D. Heitzler D, et al. 2011. "Mapping the follicle stimulating hormone induced signaling networks." *Front Endocrinol (Lausanne);* 2:45.

Halloran B. P., H. F. DeLuca. 1980. "Effect of vitamin D deficiency on fertility and reproductive capacity in the female rat." *J Nutr;* 110:1573–80.

Harkness, L. S., A. E. Bonny. 2005. "Calcium and vitamin D status in the adolescent: Key roles for bone, body weight, glucose tolerance, and estrogen biosynthesis." *J Pediatr Adolesc Gynecol;* 18:305–11.

He, C., Z. Lin, S. Robb, et al. 2015. "Serum vitamin D levels and polycystic ovary syndrome: a systematic review and meta-analysis." *Nutrients;* 7:4555–4577.

Heaney, R. P. 2008. "Vitamin D in health and disease." *Clin J Am Soc Nephrol;* 3:1535–41.

Hollis, B. W., C. L. Wagner. 2006. "Vitamin D deficiency during pregnancy: an ongoing epidemic." *Am. J. Clin. Nutr.* 84, 273.

Hsu, S. Y., R. J. Lai, M. Finegold, A. J. Hsueh. 1996. "Targeted over-expression of Bcl-2 in ovaries of transgenic mice leads to decreased follicle apoptosis, enhanced folliculogenesis, and increased germ cell tumorigenesis." *Endocrinology;* 137:4837–43.

Huijgen, N. A.; M. E. van de Kamp, J. M. Twigt, J. H. de Vries, P. H. Eilers, E. A. Steegers et al. 2014. "The preconception dietary risk score; a simple tool to assess an inadequate habitual diet for clinical practice." *e-SPEN J.* 9, e13–e19.

Hurley, W. L., R. M. Doane. 1989. "Recent developments in the roles of vitamins and minerals in reproduction." *J Dairy Sci;* 72:784–804.

International Association of Diabetes and Pregnancy Study Groups Consensus Panel, Metzger BE, S. G. Gabbe, B. Persson, et al. 2010. "International association of diabetes and pregnancy study groups recommendations on the diagnosis and classification of hyperglycemia in pregnancy." *Diabetes Care*. 33(3):676-682.

Irani, M., Merhi Z. 2014. "Role of vitamin D in ovarian physiology and its implication in reproduction: a systematic review." *Fertil Steril;* 102(2):460-468 e3.

Itoh, M. T., B. Ishizuka, Y. Kuribayashi, A. Amemiya, Y. Sumi. 1999. "Melatonin, its precursors, and synthesizing enzyme activities in the human ovary." *Mol Hum Reprod;* 5:402–8.

Jain, P., M. Jain, C. Haldar, T. B. Singh, S. Jain. 2013. "Melatonin and its correlation with testosterone in polycystic ovarian syndrome." *J Hum Reprod Sci;* 6(4):253-258.

Januszewski, M., T. Issat, et al. 2019. "Metabolic and hormonal effects of a combined Myo-inositol and d-chiro-inositol therapy on patients with polycystic ovary syndrome (PCOS)." *Ginekol Pol*; 90(1):7-10.

Javaid, M., S. Crozier, N. Harvey, C. Gale, E. Dennison, B. Boucher, et al. 2006. "Maternal vitamin D status during pregnancy and childhood bone mass at age 9 years: A longitudinal study." *Lancet,* 367, 36–43.

Johnson, L. E., H. F. DeLuca. 2001. "Vitamin D receptor null mutant mice fed high levels of calcium are fertile." *J Nutr;* 131:1787–91.

Johnson, L. E., H. F. DeLuca. 2002. "Reproductive defects are corrected in vitamin D-deficient female rats fed a high calcium, phosphorus and lactose diet." *J Nutr;* 132:2270–*3.*

Jones, G. E. S. 1949. "Some newer aspects of management of infertility." *JAMA,* 141, pp. 1123-1129.

Jones, G. E. 1949. "Some newer aspects of the management of infertility." *J Am Med Assoc;* 141:1123–19.

Jordan, J., K., Craig, D. K. Clifton, M. R. Soules. 1994. "Luteal phase defect: the sensitivity and specificity of diagnostic methods in common clinical use." *Fertil Steril;* 62:54–62.

Ju, S., G. Shaltiel, A. Shamir, et al. 2004. "Human 1-D-myo-inositol-3-phosphate synthase is functional in yeast." *J Biol Chem;* 279: 21759–65.

Jukic, A. M., A. Z. Steiner, D. D. Baird. 2015. "Lower plasma 25-hydroxyvitamin D is associated with irregular menstrual cycles in a cross-sectional study." *Reprod Biol Endocrinol;* 13:20.

Jukic, A. M., K. Upson, Q. E. Harmon, D. D. Baird. 2016. "Increasing serum 25-hydroxyvitamin D is associated with reduced odds of long menstrual cycles in a cross-sectional study of African American women." *Fertil Steril;* 106(1):172-179 e2.

Kajantie, E., D. I. Phillips. 2006. "The effects of sex and hormonal status on the physiological response to acute psychosocial stress." *Psycho-neuroendocrinology, 31,* pp. 151-178.

Khandelwal, M., E. A. Reece, Y. K. Wu, M. Borenstein. 1998. "Dietary myoinositol therapy in hyperglycemia-induced embryopathy." *Teratology.* 57(2):79-84.

Larner, J. 2001. "D-chiro-inositol in insulin action and insulin resistance-old-fashioned biochemistry still at work." *IUBMB Life;* 51: 139-148.

Larner, J., D. L. Brautigan, M. O. Thorner. 2010. "D-ChiroInositol glycans in insulin signaling and insulin resistance." *Mol Med* 16:543–52.

Lerchbaum, E., B. Obermayer-Pietsch. 2012. "Vitamin D and fertility: a systematic review." *European journal of endocrinology / European Federation of Endocrine Societies;* 166(5):765-78.

Lieberman, H. R., F. Waldhauser, G. Garfeld, H. J. Lynch, R. J. Wurtman. 1984. "Effects of melatonin on human mood and performance." *Brain Res* 323(2):201–207.

Luk, J., S. Torrealday, G. P. Neal, L. Pal. 2012. "Relevance of vitamin D in reproduction." *Hum Reprod;* 27(10):3015-27.

M. Filicori, M., C. Flamigni, M. C. Meriggiola, P. Ferrari, L. Michelacci, E. Campaniello, et al. 1991. "Endocrine response determines the clinical outcome of pulsatile gonadotropin-releasing hormone ovulation induction in different ovulatory disorders." *J Clin Endocrinol Metab,* 72, pp. 965-972.

Menichini, D., F. Facchinetti. 2019. "Effects of vitamin D supplementation in women with polycystic ovary syndrome: a review." *Gynecol Endocrinol.* Jun 12:1-5.

Mokhtari, F., F. Akbari Asbagh, O. Azmoodeh, M. Bakhtiyari, A. Almasi-Hashiani. 2019. "Effects of Melatonin Administration on Chemical Pregnancy Rates of Polycystic Ovary Syndrome Patients Undergoing Intrauterine Insemination: A Randomized Clinical Trial." *Int J Fertil Steril.;* 13(3): 225-229.

Morley, R., J. Carlin, J. Pasco, J. Wark, A. Ponsonby. 2009. "Maternal 25-hydroxyvitamin D concentration and offspring birth size: Effect modification by infant vdr genotype." *Eur. J. Clin. Nutr.* 63, 802–804.

Moszkowski, E., J. D. Woodruff, G. E. Jones. 1962. "The inadequate luteal phase." *Am J Obstet Gynecol, 83,* pp. 363-72.

Muechler, E. K., K. E. Huang, J. Zongrone. 1987. "Superovulation of habitual aborters with subtle luteal phase deficiency." *Int J Fertil, 32, pp. 359-365.*

Mulligan, M. L., S. K Felton, A. E. Riek, C. Bernal-Mizrachi. 2009. "Implications of vitamin D deficiency in pregnancy and lactation." *Am. J. Obst. Gynecol.* doi: 10.1016/j.ajog.2009.09.0002.

Mumford, S. L., A. Z. Steiner, A. Z. Pollack, N. J. Perkins, A. C. Filiberto, et al. 2012. "The utility of menstrual cycle length as an indicator of cumulative hormonal exposure." *J Clin Endocrinol Metab;* 97(10):E1 871-9.

Murayama, T., M. Kawashima, T. Takahashi, T. Yasuoka, T. Kuwayama, K. Tanaka. 1997. "Direct action of melatonin on hen ovarian granulosa cells to lower responsiveness to luteinizing hormone." *Proc Soc Exp Biol Med* 215(4):386–392.

Mustonen, A. M., P. Nieminen, H. Hyvärinen. 2001. "Preliminary evidence that pharmacologic melatonin treatment decreases rat ghrelin levels." *Endocrine* 16(1):43–46.

Nakamura, Y., H. Tamura, H. Takayama, H. Kato. 2003. "Increased endogenous level of melatonin in preovulatory human follicles does not directly influence progesterone production." *Fertil Steril;* 80:1012–6.

Nestler, J. E. 1997. "Role of hyperinsulinemia in the pathogenesis of the polycystic ovary syndrome, and its clinical implications." *Semin Reprod Endocrinol;* 15: 111-122.

Niles, L. P., J. Wang, L. Shen, D. K. Lobb, E. V. Younglai. 1999. "Melatonin receptor mRNA expression in human granulosa cells." *Mol Cell Endocrinol* 156(1–2):107–110.

Nordio, M., S. Basciani, E. Camajani. 2019. "The 40:1 myo-inositol/D-chiro-inositol plasma ratio is able to restore ovulation in PCOS patients: comparison with other ratios." *Eur Rev Med Pharmacol Sci.* Jun; 23(12):5512-5521. doi: 10.26355/eurrev_201906_18223.

Olson, J. L., R. W. Rebar, J. R. Schreiber, J. L. Vaitukaitis. 1983. "Shortened luteal phase after ovulation induction with human menopausal gonadotropin and human chorionic gonadotropin." *Fertil Steril,* 39, pp. 284-291.

Pacchiarotti, A., G. Carlomagno, G. Antonini & A. Pacchiarotti (2015): Effect of myo-inositol and melatonin versus myo-inositol, in a randomized controlled trial, for improving *in vitro* fertilization of patients with polycystic ovarian syndrome, *Gynecological Endocrinology,* doi: 10.3109/09513590.2015.1101444.

Panda, D. K., D. Miao, M. L. Tremblay, J. Sirois, R. Farookhi, G. N. Hendy, et al. 2001. "Targeted ablation of the 25-hydroxyvitamin D 1 alpha-hydroxylase enzyme: evidence for skeletal, reproductive, and immune dysfunction." *Proc Natl Acad Sci USA;* 98:7498–503.

Papaleo, E., V. Unfer, J. Baillargeon, T. Chiu. 2009. "Contribution of myo-inositol to reproduction." *Eur J Obstet Gynecol Reprod Biol*; 147: 120-123.

Papaleo, E., V. Unfer, J. P. Baillargeon, F. Fusi, F. Occhi, L. De Santis. 2009. "Myo-inositol may improve oocyte quality in intracytoplasmic sperm injection cycles. A prospective, controlled, randomized trial." *Fertil Steril;* 91: 1750-1754.

Pappolla, M. A., Y. J. Chyan, B. Poeggeler, B. Frangione, G. Wilson, J. Ghiso, et al. 2000. "An assessment of the antioxidant and the antiamyloidogenic properties of melatonin: implications for Alzheimer's disease." *J Neural Transm;* 107:203–31.

Parisi, F., A. Laoreti, I. Cetin. 2014. "Multiple micronutrient needs in pregnancy in industrialized countries." *Ann. Nutr. Metab.* 65, 13–21.

Pfeifer, S., R. Reindollar, R. Sokol, C. Gracia, R. Rebar, A. La Barbera, et al. 2015. "Current clinical irrelevance of luteal phase deficiency: a committee opinion." *Fertil Steril;103(4):e27–*Alliende, M. E., J. A. Arraztoa, U. Guajardo, F. Mellado. 2018. "Towards the clinical evaluation of the Luteal Phase in fertile women: preliminary study of normative urinary hormone profiles." *Front Public Health;* 31(6):147.

Pirke, K. M., U. Schweiger, W. Lemmel, J. C. Krieg, M. Berger. 1985. "The influence of dieting on the menstrual cycle of healthy young women." *J Clin Endocrinol Metab, 60, pp. 1174-1179.*

Prata Lima, M. F., E. C. Baracat, M. J. Simoes . 2004. "Effects of melatonin on the ovarian response to pinealectomy or continuous light in female rats: similarity with polycystic ovary syndrome." *Braz J Med Biol Res;* 37(7):987-995.

Prior, J. C. 2005. Ovarian aging and the perimenopausal transition: the paradox of endogenous ovarian hyperstimulation. *Endocr (*2005) 26: 297. https://doi.org/10.1385/ENDO:26:3:297.

Rashidi, B., F. Haghollahi, M. Shariat, F. Zayerii. 2009. "The effects of calcium-vitamin D and metformin on polycystic ovary syndrome: a pilot study." *Taiwan J Obstet Gynecol;* 48(2):142-7.

Rizzo, P., E. Raffone, V. Benedetto. 2010. "Effect of the treatment with myo-inositol plus folic acid plus melatonin in comparison with a treatment with myo-inositol plus folic acid on oocyte quality and pregnancy outcome in IVF cycles. A prospective, clinical trial." *Eur Rev Med Pharmacol Sci. Jun;* 14(6):555-61.

Ross, A. C., J. E. Manson, S. A. Abrams, J. F. Aloia, P. M. Brannon, S. K. Clinton, R. A. Durazo-Arvizu, J. C. Gallagher, et al. 2011. "The 2011 report on dietary reference intakes for calcium and vitamin D from the Institute of Medicine: what clinicians need to know." *J Clin Endocrinol Metab.* Jan; 96(1):53-8. doi: 10.1210/jc.2010-2704. Epub 2010 Nov 29.

Strott, C. A., C. M. Cargille, G. T. Ross, M. B. Lipsett. 1970. "The short luteal phase" *J Clin Endocrinol Metab,* 30, pp. 246-251.

Suh, B. Y., G. Betz. 1993. "Altered luteinizing hormone pulse frequency in early follicular phase of the menstrual cycle with luteal phase deficiency patients in women." *Fertil Steril,* 60, pp. 800-805.

Sun, T. H., D. B. Heimark, T. Nguygen, J. L. Nadler, J. Larner . 2002. "Both myo-inositol to chiro-inositol epimerase activities and chiro-inositol to myo-inositol ratios are decreased in tissues of GK type 2 diabetic rats compared to Wistar controls." *Biochem Biophys Res Commun;* 293: 1092-1098.

Sun, Th, Db Heimark, T. Nguygen, J. Nadler, J. Larner. 2002. "Both myo-inositol to chiro-inositol epimerase activities and chiro-inositol to myo-inositol ratios are decreased in tissues of GK type 2 diabetic rats compared to Wistar controls." *Biochem Biophys Res Commun*; 293: 1092-1098.

Swyer, G. I. M., D. Daley. 1952. "Progesterone implantation in habitual abortion." *BMJ,* pp. 1073-1086.

Tagliaferri, V., D. Romualdi, E. Scarinci, S. Cicco, C. D. Florio, V. Immediata, A. Tropea, C. M. Santarsiero, A. Lanzone, R. Apa. 2018. "Melatonin treatment may be able to restore menstrual cyclicity in women with PCOS: a pilot study." *Reprod Sci* 25(2):269–275.

Tamura, H., A. Takasaki, I. Miwa, K. Taniguchi, R. Maekawa, H. Asada, et al. 2008. "Oxidative stress impairs oocyte quality and melatonin protects oocytes from free radical damage and improves fertilization rate." *J Pineal Res;* 44:280–7.

Tamura, H., A. Takasaki, T. Takeani, et al. 2013. "Melatonin as a free radical scavenger in the ovarian follicle." *Endocr J.;*60(1): 1-13.

Tamura, H., Y. Nakamura, A. Korkmaz, L. C. Manchester, D. X. Tan, N. Sugino, R.J. Reiter. 2009. "Melatonin and the ovary: physiological and pathophysiological implications." *Fertil Steril* 92(1):328– 343 (Review).

Tanavde V. S., A. Maitra. 2003. "*In vitro* modulation of steroidogenesis and gene expression by melatonin: a study with porcine antral follicles." *Endocr Res.* 29(4):399-410.

Tavaniotou, A., C. Albano, J. Smitz, P. Devroey. 2002. "Impact of ovarian stimulation on corpus luteum function and embryonic implantation." *J Reprod Immunol,* 55, pp. 123-130.

Unfer, V., E. Raffone, P. Rizzo,S. Buffo. 2011. "Effect of a supplementation with myo-inositol plus melatonin on oocyte quality in women who failed to conceive in previous *in vitro* fertilization cycles for poor oocyte quality: a prospective, longitudinal, cohort study." *Gynecological Endocrinology* Early Online, 1–5.

Unfer, V., G. Porcaro. 2014. "Updates on the myoinositol plus D-chiro-inositol combined therapy in polycystic ovary syndrome." *Expert Rev Clin Pharmacol* 7:623-31.

Unfer, V., J. E. Nestler, Z. A. Kamenov, N. Prapas, & F. Facchinetti. 2016. "Effects of Inositol(s) in Women with PCOS: A Systematic Review of Randomized Controlled Trials." *International Journal of Endocrinology*, Article ID 1849162, 12 pages. http://dx.doi.org/10.1155/2016/1849162.

Van der Linden, M., K. Buckingham, C. Farquhar, J. A. M. Kremer, M. Metwally. 2015. "Luteal phase support for assisted reproduction cycles." *Cochrane Database Syst Rev;* 7:CD009154.

Viganò, P., D. Lattuada, S. Mangioni, L. Ermellino, M. Vignali, et al. 2006. "Cycling and early pregnant endometrium as a site of regulated expression of the vitamin D system." *J. Mol. Endocrinol.* 36, 415–424.

Webley, G. E., F. Leidenberger. 1986. "The circadian pattern of melatonin and its positive relationship with progesterone in women." *J Clin Endocrinol Metab* 63(2):323–328.

Wetterberg, L., J. Arendt, L. Paunier, P. C. Sizonenko, W. Donselaar, T. Heyden. 1976. "Human serum melatonin changes during the menstrual cycle." *J Clin Endocrinol Metab* 42(1):185–188.

Wetterberg, L., J. Arendt, L. Paunier, P. C. Sizonenko, W. Donselaar, T. Heyden. 1976. "Human serum melatonin changes during the menstrual cycle." *J Clin Endocrinol Metab;* 42:185–8.

Wilson, M.P., C. Hugge, M. Bielinska, P. Nicholas, P. W. Majerus, D. B. Wilson. 2009. "Neural tube defects in mice with reduced levels of inositol 1,3,4-trisphosphate 5/6-kinase." *Proc Natl Acad Sci USA.* 106 (24):9831-9835.

Woo, M. M., C. J. Tai, S. K. Kang, P. S. Nathwani, S. F. Pang, P. C. Leung. 2001. "Direct action of melatonin in human granulosa-luteal cells." *J Clin Endocrinol Metab* 86(10):4789–4797.

Yoshizawa, T., Y. Handa, Y. Uematsu, S. Takeda, K. Sekine, Y. Yoshihara, et al. 1997. "Mice lacking the vitamin D receptor exhibit impaired bone formation, uterine hypoplasia and growth retardation after weaning." *Nat Genet;* 16:391–6.

Zhang, C., C. Qiu, F. B. Hu, R. M. David, R. M. Van Dam, A. Bralley, M. A. Williams. 2008. "Maternal plasma 25-hydroxyvitamin d concentrations and the risk for gestational diabetes mellitus." *PLoS ONE,* 3, e3753.

In: Vitamin D as Progesterone-Like Hormone ISBN: 978-1-53617-516-5
Editor: Vittorio Unfer

Chapter 8

THE ROLE OF VITAMIN D IN THE MANAGEMENT OF UTERINE FIBROIDS: THE POSSIBLE SYNERGY WITH EPIGALLOCATECHIN GALLATE

Giusy Porcaro[1,*], MD and Beatrice Orrù[2], PhD
[1]Women's Health Center, Azienda USL Umbria 2, Orvieto, Italy
[2]Medical Affairs Department, Lo. Li. Pharma, Rome, Italy

ABSTRACT

Uterine fibroids (UFs) are the most common benign tumors in females. Different therapeutic approaches are available for the treatment of UFs, although, so far long-term treatments have been very limited. Treatment options include pharmacological approaches or surgery, and depend on symptomatology, size, number and desire for future pregnancy. UFs may be asymptomatic and require no intervention or further investigations, however, almost a third of women with UFs requires a therapy. Minimal invasive procedures or alternative medical treatments for handling UFs would be preferred to the traditional abdominal surgery. A number of

* Corresponding Author's Email: giusy.porcaro@gmail.com.

preclinical and clinical studies showed the potential role of vitamin D in the management of UFs and the related symptoms. Another molecule, epigallocatechin gallate (EGCG) gained attention as an optimal therapeutic approach for UFs. Therefore, vitamin D and EGCG supplementation seems very promising as a novel treatment for UFs.

Keywords: uterine myoma, uterine fibroids, leiomyoma, vitamin D, epigallocatechin gallate, EGCG

UTERINE FIBROIDS

Uterine fibroid (UF), also known as leiomyoma or myoma, is the most common benign uterine neoplasm (Sparic et al. 2016); it is a monoclonal tumor of the smooth muscle cells of the myometrium and consists of a large amount of extracellular matrix containing collagen, fibronectin and proteoglycan. UFs are typically round well-circumscribed masses.

It is known that between 5.4 to 77% of women develop UFs during their lifetime (Evans and Brunsell 2007), and about 30% are above 35 years of age. Different epidemiological studies have confirmed that UF prevalence is higher in the United States than in Europe, presumably due to racial and ethnic differences (Somigliana et al. 2007, Chibber et al. 2016). Indeed, the overall incidence of UFs is estimated to be 3-4 times higher in African American women compared to Caucasian (Chibber et al. 2016, Eltoukhi et al. 2014). The development of UFs varies greatly in relation to age, heredity, obesity, hormonal and environmental factors. Additional environmental factors, such as diet, particularly vitamin D deficiency, and environmental toxins, are the subject of ongoing investigations.

UFs are not observed in prepubescent girls and are rare in adolescents, since they are dependent on hormonal changes. According to the available data, the growth of UFs depends mostly on the influence of steroid hormones. Estrogens have been known to play an important role in their pathophysiology, but progesterone seems to be the main factor initiating uterine muscle differentiation and abnormal growth.

UFs morphology and numerousness may vary greatly. They can be solitary or appear in multiple clusters. Fibroids range in size from undetectable by the human eye, to bulky masses that can distort and enlarge the uterus. During its growth, the myometrium and connective tissue can be compressed, leading to the formation of a sort of pseudocapsule, rich in collagen fibers, neurofibers and blood vessels (Tinelli et al. 2009, Tinelli et al. 2012).

UFs are classified depending on the location in the uterine cavity: intramural fibroids, subserosal fibroids, submucosal fibroids and cervical fibroids. Intramural fibroids are the most common type and usually are asymptomatic. They originate as small nodules in the muscular wall of the uterus but may expand inwards, causing distortion and elongation of the uterine cavity. Subserosal fibroids are located on the surface of the uterus. They can be pedunculated when attached to the uterus by a small piece of tissue. Submucosal fibroids are situated in the muscle beneath the endometrium of the uterus causing bleeding and infertility. Indeed, the presence of submucosal fibroids reduces the fertility outcomes (Pritts, Parker, and Olive 2009). Therefore, the adverse impact on fertility and pregnancy encourages women for removal of submucosal fibroids. Cervical fibroids develop in the wall of the cervix and can change the shape of the cervix or cause its lengthening. They might be found rarely in the round, broad, or uterosacral ligament of the uterus. If the cervical fibroid gets extremely bigger it may obstruct the cervix, causing urinary problems and irregular menstrual cycles. This type of fibroids might be quite dangerous during pregnancy, since they can block the fetus from exiting the womb.

UFs commonly are asymptomatic, but in 30% of cases they can cause various symptoms, such as dysmenorrhea, menorrhagia, sense of weight in the pelvis, compression of the surrounding organs, infertility, recurrent abortion, preterm labor as well as anemia (Munro, Critchley, and Fraser 2011). Symptoms may vary depending on the types of UFs. These clinical complications negatively impact women's health. By the way, UFs are not associated with an increased risk of uterine tumor and very rarely develop into cancer.

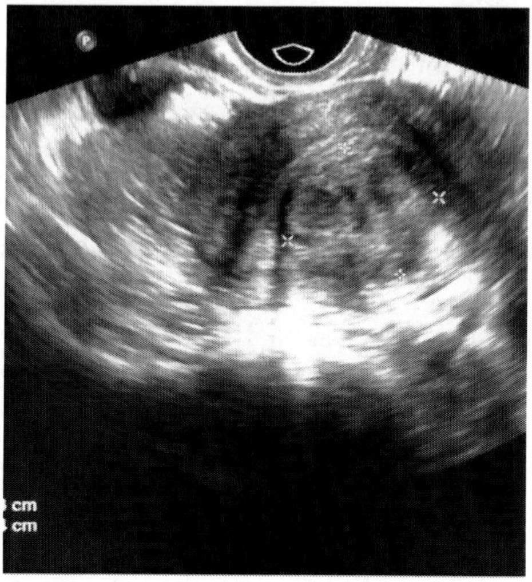

Figure 1. Intramural uterine fibroid of the uterine fundus, with clear margins, with small calcified areas inside it.

A number of studies suggested that progesterone and estrogen play a major role in UF growth and function, through their nuclear receptors (Al-Hendy et al. 2004, Rein, Barbieri, and Friedman 1995). Progesterone, progestins, and the progesterone receptor seem to modulate the fibroid mitotic activity. More precisely, progestins inhibit and/or reverse the shrinkage of UFs, which is due to hypoestrogenism induced by a gonadotropin-releasing hormone agonist (Rein, Barbieri, and Friedman 1995).

Transvaginal ultrasound is the gold standard method for the diagnosis of UF, however, occasionally abdominal ultrasound and/or magnetic resonance imaging (MRI) are performed for a major accuracy. About 3% to 7% of untreated fibroids during premenopause regress over six months to three years, and partial or complete shrinkage in menopause is normally observed. UFs may also regress during pregnancy or postpartum involution. Nevertheless, UFs affect women quality of life and are associated with significant morbidity. Therefore, there is considerable interest in developing novel techniques for the management of these benign tumors affecting the female genital tract.

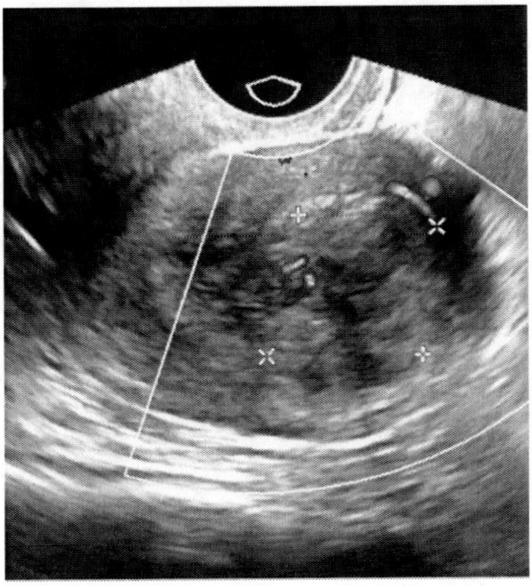

Figure 2. Intramural uterine fibroid of the uterine fundus, with areas of vascularization as per vascular pattern 2.

THERAPEUTIC APPROACHES

Nowadays, different therapeutic approaches are available for the treatment of UFs, although, so far long-term treatments of UFs have been very limited. In recent years, innovative strategies and techniques are developing for the management of UF pathology. UFs can be asymptomatic and require no intervention or further investigations ("watching and waiting"), however, almost a third of women with UFs will need a treatment, due to a variety of symptoms (Donnez and Dolmans 2016). Surgery is the most common approach in patients with symptomatic and large UFs. These include myomectomy, hysterectomy, myolysis, uterine artery embolization (UAE) and magnetic resonance imaging-guided focused ultrasound surgery (MRgFUS). The choice of the specific approach may depend on different factors such as the tumor size and characteristics, symptom severity, the pregnancy desire, and the importance of uterine preservation.

Myomectomy, known also as fibroidectomy, is the surgical removal of UFs that preserve the uterus, and therefore, the woman maintains her reproductive potential. It may be performed through an open incision, laparoscopically or through the vagina (hysterescopy).

Hysterectomy is another therapeutic approach consisting in the surgical removal of the uterus. It can be total (removing the body, fundus, and cervix of the uterus; often called "complete") or partial (removal of the uterine body while leaving the cervix intact; also called "supracervical"). Hysterectomy is the second most commonly performed gynecological surgical procedure in the United States. It is mainly adopted for the management of benign conditions such as endometriosis, irregular bleeding and UFs, and in 2015, according to the American Journal of Obstetrics and Gynecology, more than 400,000 hysterectomies were performed in the United States. Hysterectomy might cause short term and long term sequelae, and in a minor portion of operated women even mortality (between 0.4-1.1 per 1000 surgeries) (Lethaby et al. 2000).

Myolysis targets the destruction of fibroids using radiofrequency electricity, supercooled cryoprobes, and focused ultrasound monitored by real time magnetic resonance imaging. This procedure has been shown to be effective in reducing uterine bleeding and the total uterine mass (Zupi et al. 2006, Goldfarb 2008). However, the large recurrences of tumors or regrowth of UF size could be related to the partial tissue death.

Laparoscopic occlusion of the uterine arteries blocks the bilateral uterine artery by using a laparoscopic bipolar coagulation technique. However, it has no specific advantage over vaginal occlusion and clinical studies showed a lower effect in terms of UF size reduction and devascularization when compared to UAE (Hald et al. 2004). UAE consists of a minimally invasive and safe procedure for treating UFs. Usually a mass of microspheres or an embolus is injected into the uterine arteries to block the flow of blood through the vessels. This procedure reduces substantially the volume of UFs, and the heavy bleeding correlated. A Cochrane review compared UAE with other medical or surgical interventions most used for the management of UFs (Gupta et al. 2014). No significant difference among the interventions was found. A higher risk for minor complications and further incidence of surgical intervention were found in those women who underwent UAE, although hospitalization seems to be shorter compared to other techniques.

All these techniques like myomectomies, hysterectomies, and UAEs entail a psychological load and expenditure to the patient and the health care system (Mas et al. 2017).

A non-invasive procedure for removing UFs is the MRgFUS, a new technology that can be an alternative to medical and surgical interventions, with minimal risks compared to myomectomy and hysterectomy.

When the symptomatology is mild, the first-line treatment for UFs is the therapy. Among the medical options mainly prescribed can be found selective progesterone receptor modulators (sPRMs), gonadotropin-releasing hormone agonist (GnRHa), hormones such as estroprogestins or progestogens for instance the levonorgestrel-releasing intrauterine system, nonsteroidal anti-inflammatory drugs (NSAIDs), and tranexamic acid.

Ulipristal acetate (UPA), a sPRMs, offers a new medical approach for UFs. A recent multicentric study demonstrated the effectiveness and safety of UPA treatment in reducing the number of surgeries required (Giarre et al. 2019). A systematic review and meta-analysis compared UPA with placebo. The therapy improved fibroid symptoms controlling the bleeding and thus causing amenorrhea (Ghonim et al. 2019). Different UPA studies showed an embryotoxic effect in animal, however, data in humans are very limited, including the risk of abortion or teratogenicity. Several side effects including nausea, abdominal pain, emesis, dysmenorrhea, headache, dizziness, and fatigue are encountered during the treatment with UPA, and its administration during pregnancy must be excluded. Mifepristone, asoprisnil and telapristone acetate are members of the family of compound SPRMs and shown to decrease UF size and reduce uterine bleeding in a dose dependent manner. However, no clear evidence was found; also some concerns on the adverse estrogenic activity and liver toxicity were raised (Tristan et al. 2012, Bouchard, Chabbert-Buffet, and Fauser 2011, Williams et al. 2007, Spitz 2009).

GnRHa is a type of medication which affects gonadotropins and sex hormones. This treatment is commonly used to reduce volume and vascularity before myomectomy. Furthermore, they induce amenorrhoea and thus reduce the blood loss during surgery and consequently anaemia (Sinai Talaulikar, Belli, and Manyonda 2012). However, the management of fibroid diseases with GnRHa therapy is quite controversial. Indeed, large and multiple UFs treated with GnRHa showed a low regression and increase the risk of 'recurrence' of smaller fibroids. Furthermore, they impair the tissue planes resulting in complicating the surgery outcome and are not cost-effective considering also the correlated-side effects such as hot flushes and bone loss.

Levonorgestrel intrauterine device is effectively used for reducing menstrual blood flow and correlated symptoms (Senol et al. 2015). This device is placed in the uterus and slowly releases the hormone (levonorgestrel). Moderate side effects are commonly present as it is released locally in low concentration (Zapata et al. 2010). NSAIDs are commonly used for reduction of pain, in combination to estroprogestins or progestogens, which mainly exert their action on the endometrium. Alternatively, tranexamic acid is used as procoagulant drug for reduction of blood loss during menstruation, with few side effects well tolerated.

Minimal invasive procedures or alternative medical treatment for handling UFs over a traditional abdominal surgery would be preferred. Since there is a minimal concern for malignancy, in asymptomatic women surveillance of fibroids is often adopted. The number, size, and location of UFs as well as the symptoms are essential for the evaluation of potential surgical techniques. Women should be aware of all available options for the management of UF and gynecologists should discuss with each patient in order to find the most appropriate treatment.

THERAPEUTIC APPLICATION OF ORAL VITAMIN D AND EPIGALLOCATECHIN GALLATE

Vitamin D is a fat-soluble secosteroid hormone involved primarily in the regulation of calcium and bone health. However, its potential role in different areas such as immunology, oncology, dermatology, rheumatology, hematology, endocrinology, obstetrics and reproductive medicine was extensively reported (Cantorna 2006, Barrea et al. 2017, Amin et al. 1999,

Hall and Juckett 2013, Hollis 2007, Anagnostis, Karras, and Goulis 2013, S. et al. 2017, Colonese et al. 2015, Vitale, Laganà, and D'Anna 2018).

The main source of vitamin D is sunlight exposure, with the solar UVB photons absorbed by 7-dehydrocholesterol in the skin and converted to previtamin D, which then isomerizes into vitamin D_3. However, Vitamin D can be obtained also from dietary sources such as fish oils, fatty fish, fortified foods, and vitamin supplements.

The 25-hydroxyvitamin D (25-OHD) levels were inversely associated with different tumors such as breast, lung, lymphomas, and colon cancers (Liu et al. 2017, Finkelmeier et al. 2015, Ekmekcioglu, Haluza, and Kundi 2017, Atoum and Alzoughool 2017). Interestingly, case fatality was associated to lower levels of 25-OHD, however the nature of this relationship is not clear yet (Robsahm et al. 2019).

The antitumor and therapeutic effects of vitamin D on UFs has been investigated in different *in vitro* and *in vivo* animal models. Indeed, it was shown that Vitamin D inhibits the growth of human uterine leiomyoma (HuLM) cell *in vitro*, through the down-regulation of the proliferating cell nuclear antigen (PCNA), cyclin-dependent kinase 1 (CDK1) and B-cell lymphoma 2 (BCL-2), and suppresses catechol-O-methyltransferase (COMT) expression (Sharan et al. 2011). The same team reconfirmed the results with another *in vivo* experiment: the treatment with subcutaneous 1,25-dihydroxyvitamin D_3 (1,25$(OH)_2D_3$) for 3 weeks reduced UF tumor size in Eker rats. In particular, 1,25$(OH)_2D_3$ arrested the cell growth and proliferation-related genes (Pcna, cyclin D1 [Ccnd1], Myc, Cdk1, Cdk2, and Cdk4), as well as antiapoptotic genes (Bcl2 and Bcl2l1 [Bcl-x]). Furthermore, it decreased the levels of estrogen and progesterone receptors. The expression of PCNA and MKI67 were reduced while the expression of caspase 3 increased (Halder, Sharan, and Al-Hendy 2012). Another study *in vitro* demonstrated that 1,25$(OH)_2D_3$ (100 nM) inhibited the growth by 62% in both myometrial and leiomyoma cells (Blauer et al. 2009). All these molecular pathways involved in the onset and development of UFs may be useful for new thrilling breakthrough in the treatment of UFs (Brakta et al. 2015).

One recent study showed that vitamin D administration reduces the size of UF, without side effects or toxicity (Hajhashemi et al. 2019). Low vitamin D serum levels have been reported to be inversely correlated with the severity of UFs and to increase the incidence of UFs (Sabry et al. 2013).

In a recent clinical trial vitamin D_3 restored the normal vitamin D serum levels in women with hypovitaminosis D (serum level <30ng/mL), UFs (<5 cm) and mild correlated symptoms (Ciavattini et al. 2016). From all these experimental evidences it seems that vitamin D administration is an effective way to treat UF.

Epigallocatechin gallate (EGCG) is another molecule that is gaining interest in research for its potential role in the growth of UFs. EGCG is a plant compound, the most abundant catechin in tea, mainly found in green tea. Approximately 40% of the total polyphenolic mixture of green tea consists of EGCG. Onions, apples, grape, plums, hazelnuts, pecans, and carob powder may also contain small quantities of EGCG. Recently, a considerable attention of the researchers was focused on this molecule for its positive impact on health. Catechins are a group of bioflavonoids well known for their antioxidant and anti-inflammatory activity (Mukhtar and Ahmad 2000, Ohishi et al. 2016, Grzesik et al. 2018). Several researches suggest that catechins like EGCG may reduce inflammation and prevent certain chronic conditions, including heart disease, diabetes, and some cancers (Kim, Quon, and Kim 2014, Pandey and Rizvi 2009). EGCG has multiple targets and exert its action in a pleiotropic manner. Its property for improving the quality of life in patients with inflammatory diseases has been extensively studied. It seems to protect the cells from damage associated with oxidative stress, through the scavenging of reactive oxygen species and the anti-inflammatory effect of EGCG may be attributed to the downregulation of nuclear factor kappa B (NF-κB) (Ohishi et al. 2016, Arafa and Atteia 2019, Zhong et al. 2019, Wang, Fan, and Zhang 2019, Liang, Ip, and Mak 2019). Green tea and EGCG are effective without severe adverse effects; however, overdosage should be avoided, as can be toxic for the liver and cause hepatic injury.

EGCG proved to perform a very beneficial action for heart wellness by reducing the major risk factors such as high blood pressure, high cholesterol levels, and the accumulation of plaque in blood vessels (Bhardwaj and Khanna 2013, Eng, Thanikachalam, and Ramamurthy 2018, Batista Gde et al. 2009, Bogdanski et al. 2012).

A number of studies have demonstrated that EGCG injections significantly reduced inflammation, as well as the time of recovery and regeneration of neural cells in mice with spinal cord injuries. More research is needed to better understand whether EGCG may effectively prevent or treat degenerative brain diseases in humans (Machova Urdzikova et al. 2017, Renno et al. 2014).

Also, the anticancer effects and potential mechanisms of EGCG were investigated. It was demonstrated that this molecule possesses a chemopreventive potential in a number of tumors and its clinical application seems to be very promising as anticancer agent (Zan et al. 2019, Wang et al. 2019, Li et al. 2019). Therefore, the use of EGCG may offer a variety of health benefits in different districts and conditions.

In vitro studies showed that EGCG induces apoptosis and promotes cell growth arrest (Gupta et al. 2004). Furthermore, the tumor-preventive effects of green tea and catechins in different organs such as skin, lung, liver, breast, prostate, stomach, mammary gland and colon were proven through preclinical and clinical studies (Chen et al. 2008, Huo et al. 2008, Khan et al. 2006, Yang et al. 2009, Conney et al. 1999, Dreosti, Wargovich, and Yang 1997). The interest on the potential role of EGCG as an effective, safe and long-term treatment for UFs is increasing worldwide. The potential activity and the effect of EGCG were investigated on HuLM cell line (Zhang et al. 2010). Interestingly, this molecule exerted an antiproliferative action in a time and dose dependent manner, as well as an induction of apoptosis. EGCG modulated the expression of several gene through multiple signaling pathways such as BCL2, BCL2A1 and Cyclin D1 (inhibition), and CDKN1A, CDKN2B, MDM2 and tumor protein p53 (enhancement). An inhibited expression of PCNA and CDK4, as well as an increment of the proapoptotic BAX expression in a dose-dependent manner were also observed.

UFs present elevated COMT levels and the anti-COMT activity of EGCG was reported in different studies (Lu, Meng, and Yang 2003, Chen et al. 2005, Shixian et al. 2006, Salama et al. 2006).

In 2011 for the first time, a double-blinded, placebo-controlled randomized clinical trial was carried out using oral 800 mg green tea extract (45% EGCG) daily for 4 months in premenopausal women with UF ≥ 2 cm^3 (Roshdy et al. 2013). Interestingly, at the end of 4 months of supplementation with EGCG, a significant reduction by an average of 32.6% of fibroid volume was observed. Furthermore, the UF-specific symptom severity percentile score significantly decreased and the health related quality of life improved in those women receiving EGCG. Average of blood loss and hemoglobin levels between the first and last visits was also calculated. The menstrual bleeding significantly decreased in the treatment group and the levels of emoglobin increased from 11.7 to 12.4 g/dL. These results postulate the potential role of EGCG for an effective, safe non-surgical medical treatment option for UFs.

Altogether, these two molecules might be considered as novel noninvasive therapeutic agents to effectively treat UFs. Further preclinical and clinical studies are warranted to better evaluate the activities of Vitamin D and EGCG and their possible synergy as a potential alternative treatment of UFs. The physiologic effects suggest multiple mechanisms of potential benefit from the use of Vitamin D and EGCG in the treatment of these benign tumors.

Vitamin D and Epigallocatechin Gallate: Features and Safety of Use

Safety surveillance is just as important for prescription drugs as well as for dietary supplements (Jordan, Jack, and Pless 2018). When taking any particular supplement, it is important to know whether the potential benefits outweigh the risks. The right doses and duration of treatment should be carefully examined and validated. For these reasons biocompatibility and safety tests should be taken into account for any medical treatment.

Histological and laboratory testing demonstrated that Eker rats well tolerated the treatment with subcutaneous $1,25(OH)_2D_3$ at the dose of 0.5 μg/kg per day. During the entire experiment (3 weeks) were not observed signs of toxicity, necrosis, or alterations in the morphology of organs such as the liver, kidneys, lung, spleen, brain, uterus, ovaries, and brain. The model animals survived during the course of the experiment and no evidence of tissue damage or unexplained deaths were recorded (Halder, Sharan, and Al-Hendy 2012). Interestingly, the UF in Eker rat model is characteristically similar to the human fibroids, therefore it is suitable for the evaluation of potential treatments.

The adequate levels of vitamin D in humans are defined between a range 30–50 ng/mL (75–125 nmol/L), as beneficial for overall health (Pludowski et al. 2018). Although uncommon, vitamin D self-administration may be correlated to adverse effects, such as hypercalcemia and hypercalciuria, possibly caused by exceeding doses for prolonged periods. The European Food Safety Authority (EFSA) set at 100 μg/day the tolerable upper intake levels of vitamin D, for adults including pregnant and lactating women (EFSA, Products, and (NDA) 2012).

EFSA provided a scientific opinion on the safety of green tea catechins from dietary sources including food supplements (EFSA et al. 2018). Based on the available data there is no evidence of hepatotoxicity below 800 mg EGCG/day up to 12 months. An increase of serum transaminases, which is indicative of liver injury, was seen only taking doses equal or above 800 mg EGCG/day. The mean daily intake of EGCG approved for an adult population in the EU is 90–300 mg/day. In another study, safety of EGCG was tested at different doses, from 400 mg/day to 1200 mg/day, in healthy volunteers. Even the highest concentrations were well tolerated and transient mild side effects such as nausea were reported in only 3% of participants (Chow et al. 2005). Therefore, from all the clinical studies reviewed on the potential adverse effects of EGCG on the liver, the EFSA Panel concluded that there is no evidence of hepatotoxicity below the daily 800 mg consumption of this molecule. However further studies are recommended to determine safety of green tea extract and examine inter and intra species variability.

CONCLUSION

Available treatment options for UFs including surgery and medical therapy are valid and effective, but with limitations. Moreover, the observing period of UFs ("watching and waiting") might not be the best choice and an alternative would be preferred. Thus, it is quite compelling to find novel approaches. The combined administration of EGCG and vitamin D in the management of UFs has not yet been tested. Not many clinical studies were conducted so far, however, their effectiveness, when used separately, in reducing the volume of UFs as well as contrasting the symptoms correlated were demonstrated. Therefore, it would be very interesting to test the possible synergy of these two molecules for UFs treatment. This novel oral approach might have a considerable impact on the reproductive health of women. Indeed, avoiding surgical treatments allows women to maintain fertility and prevents complications at the gynecological level. An appropriate treatment may improve quality of life, in a condition where the health-related condition is considerably impaired.

Additional research is needed to validate the Vitamin D and EGCG supplementation combined and to provide a simple nonsurgical approach. New pre-clinical and clinical study are required to help understanding how these two molecules work together in women with UFs. Besides, finding the appropriate dosage of each compound is crucial to identify the best effective therapy. Up to date, Vitamin D and EGCG were found safe, however, these findings need to be confirmed. In conclusion, Vitamin D and EGCG supplementation seems very promising for becoming an optimal approach for the management of UFs and the symptoms correlated.

REFERENCES

Al-Hendy, A., E. J. Lee, H. Q. Wang, and J. A. Copland. 2004. "Gene therapy of uterine leiomyomas: adenovirus-mediated expression of dominant negative estrogen receptor inhibits tumor growth in nude mice." *Am J Obstet Gynecol* 191 (5):1621-31. doi: 10.1016/j.ajog.2004.04.022.

Amin, Shreyasee, Michael P. LaValley, Robert W. Simms, and David T. Felson. 1999. "The role of vitamin D in corticosteroid-induced osteoporosis: A meta-analytic approach." *Arthritis & Rheumatism* 42 (8):1740-1751. doi: 10.1002/1529-0131(199908)42:8<1740:AID-ANR25>3.0.CO;2-E.

Anagnostis, Panagiotis, Spyridon Karras, and Dimitrios G. Goulis. 2013. "Vitamin D in human reproduction: a narrative review." *International Journal of Clinical Practice* 67 (3):225-235. doi: 10.1111/ijcp.12031.

Arafa, M. H., and H. H. Atteia. 2019. "Protective Role of Epigallocatechin Gallate in a Rat Model of Cisplatin-Induced Cerebral Inflammation and Oxidative Damage: Impact of Modulating NF-kappaB and Nrf2." *Neurotox Res*. doi: 10.1007/s12640-019-00095-x.

Atoum, Manar, and Foad Alzoughool. 2017. "Vitamin D and Breast Cancer: Latest Evidence and Future Steps." *Breast cancer : basic and clinical research* 11:1178223417749816-1178223417749816. doi: 10.1177/1178223417749816.

Barrea, Luigi, Maria Cristina Savanelli, Carolina Di Somma, Maddalena Napolitano, Matteo Megna, Annamaria Colao, and Silvia Savastano. 2017. "Vitamin D and its role in psoriasis: An overview of the dermatologist and nutritionist." *Reviews in Endocrine and Metabolic Disorders* 18 (2):195-205. doi: 10.1007/s11154-017-9411-6.

Batista Gde, A., C. L. Cunha, M. Scartezini, R. von der Heyde, M. G. Bitencourt, and S. F. Melo. 2009. "Prospective double-blind crossover study of Camellia sinensis (green tea) in dyslipidemias." *Arq Bras Cardiol* 93 (2):128-34. doi: 10.1590/s0066-782x2009000800010.

Bhardwaj, P., and D. Khanna. 2013. "Green tea catechins: defensive role in cardiovascular disorders." *Chin J Nat Med* 11 (4):345-53. doi: 10.1016/s1875-5364(13)60051-5.

Blauer, M., P. H. Rovio, T. Ylikomi, and P. K. Heinonen. 2009. "Vitamin D inhibits myometrial and leiomyoma cell proliferation *in vitro*." *Fertil Steril* 91 (5):1919-25. doi: 10.1016/j.fertnstert.2008.02.136.

Bogdanski, P., J. Suliburska, M. Szulinska, M. Stepien, D. Pupek-Musialik, and A. Jablecka. 2012. "Green tea extract reduces blood pressure, inflammatory biomarkers, and oxidative stress and improves parameters associated with insulin resistance in obese, hypertensive patients." *Nutr Res* 32 (6):421-7. doi: 10.1016/j.nutres.2012.05.007.

Bouchard, P., N. Chabbert-Buffet, and B. C. Fauser. 2011. "Selective progesterone receptor modulators in reproductive medicine: pharmacology, clinical efficacy and safety." *Fertil Steril* 96 (5):1175-89. doi: 10.1016/j.fertnstert.2011.08.021.

Brakta, S., J. S. Diamond, A. Al-Hendy, M. P. Diamond, and S. K. Halder. 2015. "Role of vitamin D in uterine fibroid biology." *Fertil Steril* 104 (3):698-706. doi: 10.1016/j.fertnstert.2015.05.031.

Cantorna, M. T. 2006. "Vitamin D and its role in immunology: multiple sclerosis, and inflammatory bowel disease." *Prog Biophys Mol Biol* 92 (1):60-4. doi: 10.1016/j.pbiomolbio.2006.02.020.

Chen, D., V. Milacic, M. S. Chen, S. B. Wan, W. H. Lam, C. Huo, K. R. Landis-Piwowar, Q. C. Cui, A. Wali, T. H. Chan, and Q. P. Dou. 2008. "Tea polyphenols, their biological effects and potential molecular targets." *Histol Histopathol* 23 (4):487-96. doi: 10.14670/hh-23.487.

Chen, D., C. Y. Wang, J. D. Lambert, N. Ai, W. J. Welsh, and C. S. Yang. 2005. "Inhibition of human liver catechol-O-methyltransferase by tea catechins and their metabolites: structure-activity relationship and molecular-modeling studies." *Biochem Pharmacol* 69 (10):1523-31. doi: 10.1016/j.bcp.2005.01.024.

Chibber, S., G. Mendoza, L. Cohen, and E. E. Marsh. 2016. "Racial and ethnic differences in uterine fibroid prevalence in a diverse cohort of young asymptomatic women (18-30 yo)." *Fertility and Sterility* 106 (3):e97. doi: 10.1016/j.fertnstert.2016.07.281.

Ciavattini, A., G. Delli Carpini, M. Serri, A. Vignini, J. Sabbatinelli, A. Tozzi, A. Aggiusti, and N. Clemente. 2016. "Hypovitaminosis D and "small burden" uterine fibroids: Opportunity for a vitamin D supplementation." *Medicine (Baltimore)* 95 (52):e5698. doi: 10.1097/md.0000000000005698.

Colonese, Francesca, Antonio Simone Laganà, Elisabetta Colonese, Vincenza Sofo, Francesca Maria Salmeri, Roberta Granese, and Onofrio Triolo. 2015. "The pleiotropic effects of vitamin D in gynaecological and obstetric diseases: an overview on a hot topic." *BioMed research international* 2015.

Conney, A. H., Y. Lu, Y. Lou, J. Xie, and M. Huang. 1999. "Inhibitory effect of green and black tea on tumor growth." *Proc Soc Exp Biol Med* 220 (4):229-33. doi: 10.1046/j.1525-1373.1999.d01-39.x.

Donnez, J., and M. M. Dolmans. 2016. "Uterine fibroid management: from the present to the future." *Hum Reprod Update* 22 (6):665-686. doi: 10.1093/humupd/dmw023.

Dreosti, I. E., M. J. Wargovich, and C. S. Yang. 1997. "Inhibition of carcinogenesis by tea: the evidence from experimental studies." *Crit Rev Food Sci Nutr* 37 (8):761-70. doi: 10.1080/10408399709527801.

EFSA, Panel on Food Additives and Nutrient Sources added to Food (ANS), Peter Aggett Maged Younes, Fernando Aguilar, Riccardo Crebelli, Birgit Dusemund, Maria Jose Frutos Metka Filipic, Pierre Galtier, David Gott, Ursula Gundert-Remy, Jean-Charles Leblanc Claude Lambré, Inger Therese Lillegaard, Peter Moldeus, Agneta Oskarsson Alicja Mortensen, Ivan Stankovic, Ine Waalkens-Berendsen, Raul J Andrade Rudolf Antonius Woutersen, Cristina Fortes, Pasquale Mosesso, and Davide Arcella Patrizia Restani, Fabiola Pizzo, Camilla Smeraldi and Matthew Wright. 2018. "Scientific opinion on the safety of green tea catechins." *EFSA Journal* 16 (4):5239, 89 pp. doi: 10.2903/j.efsa.2018.5239.

EFSA, Panel on Dietetic Products, and Nutrition and Allergies (NDA). 2012. "Scientific Opinion on the Tolerable Upper Intake Level of vitamin D." *EFSA Journal* 10 (7):2813. 45 pp. doi: 0.2903/j.efsa.2012.2813.

Ekmekcioglu, Cem, Daniela Haluza, and Michael Kundi. 2017. "25-Hydroxyvitamin D Status and Risk for Colorectal Cancer and Type 2 Diabetes Mellitus: A Systematic Review and Meta-Analysis of Epidemiological Studies." *International journal of environmental research and public health* 14 (2):127. doi: 10.3390/ijerph14020127.

Eltoukhi, Heba M., Monica N. Modi, Meredith Weston, Alicia Y. Armstrong, and Elizabeth A. Stewart. 2014. "The health disparities of uterine fibroid tumors for African American women: a public health issue." *American journal of obstetrics and gynecology* 210 (3):194-199. doi: 10.1016/j.ajog.2013.08.008.

Eng, Q. Y., P. V. Thanikachalam, and S. Ramamurthy. 2018. "Molecular understanding of Epigallocatechin gallate (EGCG) in cardiovascular and metabolic diseases." *J Ethnopharmacol* 210:296-310. doi: 10.1016/j.jep.2017.08.035.

Evans, Patricia, and Susan Brunsell. 2007. "Uterine fibroid tumors: diagnosis and treatment." *American family physician* 75 (10).

Finkelmeier, Fabian, Bernd Kronenberger, Stefan Zeuzem, Albrecht Piiper, and Oliver Waidmann. 2015. "Low 25-Hydroxyvitamin D Levels Are Associated with Infections and Mortality in Patients with Cirrhosis." *PloS one* 10 (6):e0132119-e0132119. doi: 10.1371/journal.pone.0132119.

Ghonim, M., R. Magdy, M. Sabbour, M. Ghonim, and A. Nabhan. 2019. "A systematic review and meta-analysis of ulipristal acetate for symptomatic uterine fibroids." *Int J Gynaecol Obstet* 146 (2):141-148. doi: 10.1002/ijgo.12868.

Giarre, G., M. Franchini, E. Castellacci, M. E. Malune, A. Di Spiezio Sardo, G. Saccone, and S. Angioni. 2019. "Ulipristal acetate in symptomatic uterine fibroids. A real-world experience in a multicentric Italian study." *Gynecol Endocrinol*: 1-4. doi: 10.1080/09513590.2019.1648419.

Goldfarb, Herbert A. 2008. "Myolysis revisited." *JSLS: Journal of the Society of Laparoendoscopic Surgeons* 12 (4):426-430.

Grzesik, M., K. Naparlo, G. Bartosz, and I. Sadowska-Bartosz. 2018. "Antioxidant properties of catechins: Comparison with other antioxidants." *Food Chem* 241:480-492. doi: 10.1016/j.foodchem. 2017.08.117.

Gupta, J. K., A. Sinha, M. A. Lumsden, and M. Hickey. 2014. "Uterine artery embolization for symptomatic uterine fibroids." *Cochrane Database of Systematic Reviews* (12). doi: 10.1002/14651858.CD0050 73.pub4.

Gupta, S., K. Hastak, F. Afaq, N. Ahmad, and H. Mukhtar. 2004. "Essential role of caspases in epigallocatechin-3-gallate-mediated inhibition of nuclear factor kappa B and induction of apoptosis." *Oncogene* 23 (14):2507-22. doi: 10.1038/sj.onc.1207353.

Hajhashemi, Maryam, Maryam Ansari, Fedyeh Haghollahi, and Bita Eslami. 2019. "The effect of vitamin D supplementation on the size of uterine leiomyoma in women with vitamin D deficiency." *Caspian Journal of Internal Medicine* 10 (2):125-131. doi: 10.22088/cjim.10. 2.125.

Hald, K., A. Langebrekke, N. E. Klow, H. J. Noreng, A. B. Berge, and O. Istre. 2004. "Laparoscopic occlusion of uterine vessels for the treatment of symptomatic fibroids: Initial experience and comparison to uterine artery embolization." *Am J Obstet Gynecol* 190 (1):37-43. doi: 10.1016/s0002-9378(03)00910-4.

Halder, S. K., C. Sharan, and A. Al-Hendy. 2012. "1,25-dihydroxyvitamin D3 treatment shrinks uterine leiomyoma tumors in the Eker rat model." *Biol Reprod* 86 (4):116. doi: 10.1095/biolreprod.111.098145.

Hall, A., .M. Juckett. 2013. "The role of vitamin D in hematologic disease and stem cell transplantation." *Nutrients* 5 (6):2206-2221.

Hollis, B. W. 2007. "Vitamin D Requirement During Pregnancy and Lactation." *Journal of Bone and Mineral Research* 22 (S2):V39-V44. doi: 10.1359/jbmr.07s215.

Huo, C., S. B. Wan, W. H. Lam, L. Li, Z. Wang, K. R. Landis-Piwowar, D. Chen, Q. P. Dou, and T. H. Chan. 2008. "The challenge of developing green tea polyphenols as therapeutic agents." *Inflammopharmacology* 16 (5):248-252.

Jordan, S. A., S. A. Jack, and R. P. Pless. 2018. "Safety Surveillance of Dietary Supplements: Importance, Challenges, and New Horizons." *Clin Pharmacol Ther* 104 (3):432-434. doi: 10.1002/cpt.1101.

Khan, N., F. Afaq, M. Saleem, N. Ahmad, and H. Mukhtar. 2006. "Targeting multiple signaling pathways by green tea polyphenol (-)-epigallocatechin-3-gallate." *Cancer Res* 66 (5):2500-5. doi: 10.1158/0008-5472.can-05-3636.

Kim, H. S., M. J. Quon, and J. A. Kim. 2014. "New insights into the mechanisms of polyphenols beyond antioxidant properties; lessons from the green tea polyphenol, epigallocatechin 3-gallate." *Redox Biol* 2:187-95. doi: 10.1016/j.redox.2013.12.022.

Lethaby, A., S. Shepperd, I. Cooke, and C. Farquhar. 2000. "Endometrial resection and ablation versus hysterectomy for heavy menstrual bleeding." *Cochrane Database Syst Rev* (2):Cd000329. doi: 10.1002/14651858.cd000329.

Li, T., N. Zhao, J. Lu, Q. Zhu, X. Liu, F. Hao, and X. Jiao. 2019. "Epigallocatechin gallate (EGCG) suppresses epithelial-Mesenchymal transition (EMT) and invasion in anaplastic thyroid carcinoma cells through blocking of TGF-beta1/Smad signaling pathways." *Bioengineered* 10 (1):282-291. doi: 10.1080/21655979.2019.1632669.

Liang, Y., M. S. M. Ip, and J. C. W. Mak. 2019. "(-)-Epigallocatechin-3-gallate suppresses cigarette smoke-induced inflammation in human cardiomyocytes via ROS-mediated MAPK and NF-kappaB pathways." *Phytomedicine* 58:152768. doi: 10.1016/j.phymed.2018.11.028.

Liu, J., Y. Dong, C. Lu, Y. Wang, L. Peng, M. Jiang, Y. Tang, and Q. Zhao. 2017. "Meta-analysis of the correlation between vitamin D and lung cancer risk and outcomes." *Oncotarget* 8 (46):81040-81051. doi: 10.18632/oncotarget.18766.

Lu, H., X. Meng, and C. S. Yang. 2003. "Enzymology of methylation of tea catechins and inhibition of catechol-O-methyltransferase by (-)-epigallocatechin gallate." *Drug Metab Dispos* 31 (5):572-9. doi: 10.1124/dmd.31.5.572.

Machova Urdzikova, L., J. Ruzicka, K. Karova, A. Kloudova, B. Svobodova, A. Amin, J. Dubisova, M. Schmidt, S. Kubinova, M.

Jhanwar-Uniyal, and P. Jendelova. 2017. "A green tea polyphenol epigallocatechin-3-gallate enhances neuroregeneration after spinal cord injury by altering levels of inflammatory cytokines." *Neuropharmacology* 126:213-223. doi: 10.1016/j.neuropharm.2017.09.006.

Mas, Aymara, Marta Tarazona, Joana Dasí Carrasco, Gloria Estaca, Ignacio Cristóbal, and Javier Monleón. 2017. "Updated approaches for management of uterine fibroids." *International journal of women's health* 9:607-617. doi: 10.2147/IJWH.S138982.

Mukhtar, H., and N. Ahmad. 2000. "Tea polyphenols: prevention of cancer and optimizing health." *Am J Clin Nutr* 71 (6 Suppl):1698S-702S; discussion 1703S-4S. doi: 10.1093/ajcn/71.6.1698S.

Munro, Malcolm, H. O. Critchley, and I. S. Fraser. 2011. *FIGO Menstrual Disorders Working Group. The FIGO classification of causes of abnormal uterine bleeding in the reproductive years*. Vol. 95.

Ohishi, T., S. Goto, P. Monira, M. Isemura, and Y. Nakamura. 2016. "Anti-inflammatory Action of Green Tea." *Antiinflamm Antiallergy Agents Med Chem* 15 (2):74-90. doi: 10.2174/187152301566616091515 4443.

Pandey, Kanti Bhooshan, and Syed Ibrahim Rizvi. 2009. "Plant polyphenols as dietary antioxidants in human health and disease." *Oxidative medicine and cellular longevity* 2 (5):270-278. doi: 10.4161/oxim.2.5.9498.

Pludowski, Pawel, Michael F. Holick, William B. Grant, Jerzy Konstantynowicz, Mario R. Mascarenhas, Afrozul Haq, Vladyslav Povoroznyuk, Nataliya Balatska, Ana Paula Barbosa, Tatiana Karonova, Ema Rudenka, Waldemar Misiorowski, Irina Zakharova, Alena Rudenka, Jacek Łukaszkiewicz, Ewa Marcinowska-Suchowierska, Natalia Łaszcz, Pawel Abramowicz, Harjit P. Bhattoa, and Sunil J. Wimalawansa. 2018. "Vitamin D supplementation guidelines." *The Journal of Steroid Biochemistry and Molecular Biology* 175:125-135. doi: https://doi.org/10.1016/j.jsbmb.2017.01.021.

Rein, M. S., R. L. Barbieri, and A. J. Friedman. 1995. "Progesterone: a critical role in the pathogenesis of uterine myomas." *Am J Obstet Gynecol* 172 (1 Pt 1):14-8. doi: 10.1016/0002-9378(95)90077-2.

Renno, W. M., G. Al-Khaledi, A. Mousa, S. M. Karam, H. Abul, and S. Asfar. 2014. "(-)-Epigallocatechin-3-gallate (EGCG) modulates neurological function when intravenously infused in acute and, chronically injured spinal cord of adult rats." *Neuropharmacology* 77:100-19. doi: 10.1016/j.neuropharm.2013.09.013.

Robsahm, T. E., S. Tretli, P. A. Torjesen, R. Babigumira, and G. G. Schwartz. 2019. "Serum 25-hydroxyvitamin D levels predict cancer survival: a prospective cohort with measurements prior to and at the time of cancer diagnosis." *Clin Epidemiol* 11:695-705. doi: 10.2147/clep.s 207230.

Laganà A. S., Vitale S. G., Ban Frangež H., Vrtačnik-Bokal E., and D'Anna R. 2017. "Vitamin D in human reproduction: the more, the better? An evidence-based critical appraisal." *Eur Rev Med Pharmacol Sci* 21 (18):4243-4251.

Sabry, M., S. K. Halder, A. S. Allah, E. Roshdy, V. Rajaratnam, and A. Al-Hendy. 2013. "Serum vitamin D3 level inversely correlates with uterine fibroid volume in different ethnic groups: a cross-sectional observational study." *Int J Womens Health* 5:93-100. doi: 10.2147/ijw h.s38800.

Salama, S. A., S. L. Ho, H. Q. Wang, J. Tenhunen, C. Tilgmann, and A. Al-Hendy. 2006. "Hormonal regulation of catechol-O-methyl transferase activity in women with uterine leiomyomas." *Fertil Steril* 86 (1):259-62. doi: 10.1016/j.fertnstert.2005.12.049.

Senol, T., I. Kahramanoglu, Y. Dogan, M. Baktiroglu, A. Karateke, and N. Suer. 2015. "Levonorgestrel-releasing intrauterine device use as an alternative to surgical therapy for uterine leiomyoma." *Clin Exp Obstet Gynecol* 42 (2):224-7.

Sharan, Chakradhari, Sunil K. Halder, Chandrasekhar Thota, Tarannum Jaleel, Sangeeta Nair, and Ayman Al-Hendy. 2011. "Vitamin D inhibits proliferation of human uterine leiomyoma cells via catechol-O-methyltransferase." *Fertility and sterility* 95 (1):247-253. doi: 10.1016/j.fertnstert.2010.07.1041.

Shixian, Q., B. VanCrey, J. Shi, Y. Kakuda, and Y. Jiang. 2006. "Green tea extract thermogenesis-induced weight loss by epigallocatechin gallate inhibition of catechol-O-methyltransferase." *J Med Food* 9 (4):451-8. doi: 10.1089/jmf.2006.9.451.

Sinai Talaulikar, Vikram, Anna-Maria Belli, and Isaac Manyonda. 2012. "GnRH agonists: do they have a place in the modern management of fibroid disease?" *Journal of obstetrics and gynaecology of India* 62 (5):506-510. doi: 10.1007/s13224-012-0206-0.

Somigliana, E., P. Vercellini, R. Daguati, R. Pasin, O. De Giorgi, and P. G. Crosignani. 2007. "Fibroids and female reproduction: a critical analysis of the evidence." *Hum Reprod Update* 13 (5):465-76. doi: 10.1093/humupd/dmm013.

Sparic, Radmila, Ljiljana Mirkovic, Antonio Malvasi, and Andrea Tinelli. 2016. "Epidemiology of Uterine Myomas: A Review." *International journal of fertility & sterility* 9 (4):424-435. doi: 10.22074/ijfs.2015.4599.

Spitz, I. M. 2009. "Clinical utility of progesterone receptor modulators and their effect on the endometrium." *Curr Opin Obstet Gynecol* 21 (4):318-24. doi: 10.1097/GCO.0b013e32832e07e8.

Tinelli, A., B. S. Hurst, G. Hudelist, D. A. Tsin, M. Stark, L. Mettler, M. Guido, and A. Malvasi. 2012. "Laparoscopic myomectomy focusing on the myoma pseudocapsule: technical and outcome reports." *Hum Reprod* 27 (2):427-35. doi: 10.1093/humrep/der369.

Tinelli, A., A. Malvasi, S. Rahimi, R. Negro, C. Cavallotti, D. Vergara, G. Vittori, and L. Mettler. 2009. "Myoma pseudocapsule: a distinct endocrino-anatomical entity in gynecological surgery." *Gynecol Endocrinol* 25 (10):661-7. doi: 10.1080/09513590903015502.

Tristan, M., L. J. Orozco, A. Steed, A. Ramírez-Morera, and P. Stone. 2012. "Mifepristone for uterine fibroids." *Cochrane Database of Systematic Reviews* (8). doi: 10.1002/14651858.CD007687.pub2.

Vitale, S. G., A. S. Laganà, and R. D'Anna. 2018. "Does Vitamin D supplementation always play a beneficial role in human reproduction? Steps towards an evidence-based management." *IJMDAT* 1 (1):e114.

Wang, J., S. M. Fan, and J. Zhang. 2019. "Epigallocatechin-3-gallate ameliorates lipopolysaccharide-induced acute lung injury by suppression of TLR4/NF-kappaB signaling activation." *Braz J Med Biol Res* 52 (7):e8092. doi: 10.1590/1414-431x20198092.

Wang, R., J. Huang, J. Chen, M. Yang, H. Wang, H. Qiao, Z. Chen, L. Hu, L. Di, and J. Li. 2019. "Enhanced anti-colon cancer efficacy of 5-fluorouracil by epigallocatechin-3- gallate co-loaded in wheat germ agglutinin-conjugated nanoparticles." *Nanomedicine* 21:102068. doi: 10.1016/j.nano.2019.102068.

Williams, A. R., H. O. Critchley, J. Osei, S. Ingamells, I. T. Cameron, C. Han, and K. Chwalisz. 2007. "The effects of the selective progesterone receptor modulator asoprisnil on the morphology of uterine tissues after 3 months treatment in patients with symptomatic uterine leiomyomata." *Hum Reprod* 22 (6):1696-704. doi: 10.1093/humrep/dem026.

Yang, C. S., X. Wang, G. Lu, and S. C. Picinich. 2009. "Cancer prevention by tea: animal studies, molecular mechanisms and human relevance." *Nat Rev Cancer* 9 (6):429-39. doi: 10.1038/nrc2641.

Zan, L., Q. Chen, L. Zhang, and X. Li. 2019. "Epigallocatechin gallate (EGCG) suppresses growth and tumorigenicity in breast cancer cells by downregulation of miR-25." *Bioengineered* 10 (1):374-382. doi: 10.10 80/21655979.2019.1657327.

Zapata, Lauren B., Maura K. Whiteman, Naomi K. Tepper, Denise J. Jamieson, Polly A. Marchbanks, and Kathryn M. Curtis. 2010. "Intrauterine device use among women with uterine fibroids: a systematic review." *Contraception* 82 (1):41-55. doi: 10.1016/j.contrace ption.2010.02.011.

Zhang, D., M. Al-Hendy, G. Richard-Davis, V. Montgomery-Rice, V. Rajaratnam, and A. Al-Hendy. 2010. "Antiproliferative and proapoptotic effects of epigallocatechin gallate on human leiomyoma cells." *Fertil Steril* 94 (5):1887-93. doi: 10.1016/j.fertnstert.2009.08.0 65.

Zhong, X., M. Liu, W. Yao, K. Du, M. He, X. Jin, L. Jiao, G. Ma, B. Wei, and M. Wei. 2019. "Epigallocatechin-3-Gallate Attenuates Microglial Inflammation and Neurotoxicity by Suppressing the Activation of Canonical and Noncanonical Inflammasome via TLR4/NF-kappaB Pathway." *Mol Nutr Food Res*:e1801230. doi: 10.1002/mnfr.201801230.

Zupi, E., M. Sbracia, D. Marconi, and M. G. Munro. 2006. "Myolysis of uterine fibroids: is there a role?" *Clin Obstet Gynecol* 49 (4):821-33. doi: 10.1097/01.grf.0000211961.91616.78.

In: Vitamin D as Progesterone-Like Hormone ISBN: 978-1-53617-516-5
Editor: Vittorio Unfer

Chapter 9

VITAMIN D AND CANCER

Noemi Monti[1,2], Sara Proietti[1], PhD, Alessandra Cucina[2,3], PhD and Mariano Bizzarri[1,*], MD, PhD

[1]Department of Experimental Medicine, Sapienza University of Rome, Systems Biology Group Lab, Rome, Italy

[2]Department of Surgery "P. Valdoni" Sapienza University of Rome, Rome, Italy

[3]Azienda Policlinico Umberto I, Rome, Italy

ABSTRACT

Vitamin D is not really a vitamin but the precursor of a steroid hormone, the calcitriol. Calcitriol is the biologically active form of this group of pro-hormones. Calcitriol binds to vitamin D receptor (VDR) to regulate various biological processes. It plays an important role in calcium homeostasis and normal bone mineralization. However, many epidemiological and observational studies show the role of Vitamin D in several diseases, such as cancer. In fact, Vitamin D deficiency increases the risk of developing cancer. There are various mechanisms by which

* Corresponding Author's Email: mariano.bizzarri@uniroma1.it.

vitamin D influences the natural history of cancer. These include the role of vitamin D in the induction of apoptosis, stimulation of cell differentiation, anti-inflammatory and antiproliferative effects and inhibition of angiogenesis, invasion and metastasis. These calcitriol actions provide a basis for its potential use in cancer therapy and chemoprevention.

Keywords: Vitamin D, 25(OH)D_3, cancer, Vitamin D supplementation

INTRODUCTION

Vitamin D is traditionally known as an essential nutrient but is a group of fat-soluble pro-hormones and their respective metabolites. Vitamin D can be synthesized in adequate amounts in the skin using the energy of ultraviolet B (UVB, 290-320 nm) radiation in sunlight (Holick, 2006). Indeed, most foods contain Vitamin D, so humans can also take this molecule from diet. The biologically active metabolite of this group is calcitriol, produced by two steps of hydroxylation of the precursor Vitamin D_3 in specific organs in the body, which binds to specific receptor to enable its physiological functions. Vitamin D is the major regulator of calcium homeostasis and normal bone mineralization in the body (Holick, 2006). However, in the last years it became clear that Vitamin D plays a non-calcemic effect but modulates also other biological functions, such as cancer. Several evidences support the relationship between vitamin D and cancer: (1) low circulating levels of vitamin D relates with increased risk of developing cancer; (2) a high intake of vitamin D is associated with a reduced risk of cancer; (3) the aggressiveness of a cancer is lower in summer when the production of vitamin D is higher; (4) polymorphisms of genes encoding proteins involved in the signal pathway of vitamin D affect the risk of developing cancer (Vuolo et al. 2012).

BIOCHEMISTRY AND CANONICAL METABOLIC PATHWAY

Vitamin D is more than a micronutrient and not technically a vitamin; rather, it is the essential precursor of a steroid hormone, calcitriol, also known as 1,25-dihydroxy-Vitamin D3 (1,25(OH)2D3), which mediates several actions in many tissues. There are two main isoforms of Vitamin D: Vitamin D2 (ergocalciferol) synthesized from ergosterol in plants (Hohman et al., 2011), and vitamin D3 (cholecalciferol), synthesized in human and the metabolite more active (Severo et al., 2009). In the skin, the energy from ultraviolet rays converts the substrate 7-dehydrocholesterol to pre-vitamin D3, which is immediately converted to vitamin D3 in a heat-dependent process. Excessive exposure to sunlight degrades previtamin-D3 and Vitamin D3 into inactive photoproducts. Vitamin D (D2 or D3), synthesized in the skin or ingested from the diet, can be stored in and then released from fat cells because its fat-soluble property (Glerup et al., 2000). In order to reach the most active form, Vitamin D3 or cholecalciferol undergoes to two sequential hydroxylation steps mediated by two P450 cytochrome enzymes in the liver and in the kidney (Jones et al., 2014). Vitamin D goes into dermal capillaries and in the blood is preferentially bound to vitamin D-binding protein (VDBP), which transports it into the liver where the first step occurs. The first hydroxylation is catalyzed by the enzyme vitamin D-25hydroxylase (predominantly CYP2R1) to yield 25-hydroxyvitamin D3 (25(OH)D3), also known as calcidiol) (Zhu et al., 2012). This is the major circulating form and it is used to determine and control the patient's vitamin D status (Ponchon et al., 1969). The second hydroxylation occurs in the kidney, where 25(OH)D3 is hydroxylated at the C1α position by the cytochrome P450 enzyme 1α-hydroxylase (CYP27B1) to produce calcitriol, the active metabolite (Fraser et al., 1970; Jones et al., 2014).

Vitamin D Measurement

Vitamin 25(OH)D3 is the circulating form of the hormone that is measured in the blood. This value is clinically used to establish the vitamin D status of a patient (Holick, 1990). There is no consensus on optimal levels of 25-hydroxyvitamin D as measured in serum, indeed, two different committees expressed their opinions, but their reports disagree. The IOM (Institute of Medicine) set the cut-points for deficiency at 20 ng per mL (50 nmol per L) and the daily requirement at 600 International units (IU) vitamin D per day for people under the age of 70, and 800 IU daily for those aged 70 years and above, mostly focusing on the vitamin D requirement for skeletal health (Holick et al., 2011). At the opposite, the Endocrine Society establish the cut-point at 30 ng per mL (75 nmol per L) and recommend higher daily doses (perhaps 1,000–2,000 IU) for many individuals and much higher doses in some individuals to maintain adequate circulating concentrations of 25(OH)D3. This controversy is based on different interpretations of what is the adequate circulating concentrations of 25(OH)D3 for bone health. The IOM recommends the usage of vitamin D for only normal bone mineralization claiming that Vitamin D benefits in the other extra-skeletal conditions, including cancer, could not be demonstrated without any concordant results obtained only through randomized clinical trials. Like the IOM, the Endocrine Society also recommends vitamin D supplementation to avoid deficiencies and to prevent falls and fractures but not to prevent cardiovascular disease, cancer or death, or to improve quality of life. Thus, the optimal range is between 30 ng/mL and 60 ng/mL. Several factors affect the vitamin D status: sunlight exposure, diet, obesity, low physical activity, lifestyle, age, skin type, smoking and living at high latitude. Thus, Vitamin D synthesis and serum 25(OH)D3 levels are inversely correlated with latitude and positively correlated with sunlight (McKenna, 1992). UVB exposure and dietary vitamin D intake increase the serum 25(OH)D3 levels in a dose-dependent manner providing a higher concentration of 25(OH)D3 used as substrate for the synthesis of 1,25(OH)2D3 (Adams et al., 1982; Holick, 1987; Webb et al., 1990). 1,25(OH)2D3 is the most active vitamin D metabolite, although its serum concentration is one thousand that of

25(OH)D3 (Holick, 1995). Since 1,25(OH)2D3 is synthesized in several tissues, this provides a possible explanation for lower incidence rates of cancer in subjects with higher levels of serum 25(OH)D3. Moreover, sunlight also increases 25(OH)D3 levels providing more substrate to synthetize 1,25(OH)2D3 and this explains the association between the degree of UVB exposure with lower mortality rates from cancer. Due to the different sun exposures, lifestyle and foods low in vitamin D, not everyone has adequate concentration in the blood and needs supplementation (Chapuy et al., 1997). The intake of vitamin D in adults age 19 years and older at risk of deficiency is recommended to be above 600 IU to 800 IU daily (National Academy of Sciences, 1997). Some individuals may require vitamin D supplements of 2000 IU per day or more to achieve optimal levels. (Vitamin D3 1000 IU daily will raise blood levels, on average, by approximately 10 ng/mL). For adults age 50 years or older, the National Osteoporosis Foundation recommends 800 IU to 1000 IU of vitamin D per day supplementation.

Molecular Biology of Vitamin D

Calcitriol exerts its nuclear effects by binding to the specific receptor, VDR, which is a member of the steroid–thyroid–retinoid receptor superfamily of ligand-activated transcription factors (Debb et al., 2007; Christakos et al., 2016). Vitamin D receptor is present in most cells in the body, and calcitriol, directly or indirectly, regulates from 3% to 5% of the human genome (De Luca et al., 2004; Wang et al., 2012). Vitamin D, bound to VDR, induces both genomic and non-genomic regulation of downstream gene targets, involving several and different biological functions (Bouillon et al., 2008; Zanatta et al., 2011). Non-genomic pathway is fast and does not depend on transcription but may interact with the genomic one thus affecting indirectly the transcription via crosstalk with other signaling pathway. In the non-genomic mechanism of action, calcitriol binds to non-classical membrane receptor: 1,25D-MARRS (1,25D-membrane-associated, rapid response steroid-binding protein). This interaction induces changes in cell signaling pathways, including the rapid translocation of calcium across intestinal mucosal membranes (Nemere et al., 2004; Norman, 2006). Binding of 1,25(OH)2D at plasma membrane may activate one or more second-messenger signaling, such as mitogen-activated protein kinase (MAPK) or Phospholipase C (PKC) or Phosphatidylinositol-3-kinase (PI3K) ones. These second messengers, particularly MAPK, may modulate a crosstalk with the genomic pathway to regulate gene expression (Morelli et al., 2001).

In the genomic pathway, calcitriol binds to VDR in the cytoplasm and this binding promotes phosphorylation of receptor and its heterodimerization with Retinoid-X-Receptor (RXR) and, moreover, the nuclear translocation of the complex (Haussler et al., 1998; Christakos et al., 2003). The Calcitriol-VDR-RXR complex recognizes and binds to VDR Element (VDRE) on promoter region of target genes recruiting transcriptional co-activators or co-repressors to regulate their expression (Carlberg et al., 1993; Pike et al., 2012). Vitamin D activity is widespread.

The bioavailability and metabolism of 1,25(OH)2D3 are tightly regulated through mechanisms of negative and positive feedbacks that change the expression of the enzymes playing a role in the calcitriol metabolism (Henry et al., 1984; DeLuca, 2008). The intracellular concentrations of calcitriol and its metabolites are determined by two key enzymes: CYP27B1 and CYP24A1. CYP27B1 is the rate-limiting enzyme in calcitriol metabolism, mainly expressed in the kidney but also in other cells. The renal CYP27B1 is crucial for the concentration of circulating calcitriol, which acts on other tissues through an endocrine pathway (Hobaus et al., 2013). In the kidney, the activity of CYP27B1 is regulated by parathyroid hormone (PTH), a calciotropic hormone, the level of which is inversely proportional to serum calcium concentration (Dusso et al., 2011; Martin et al., 2012) (Boyle et al., 1971; Henry, 1985; Murayama et al., 1999). Low calcium concentration and high PTH levels stimulate the transcription of CYP27B1 in the kidney promoting the calcitriol synthesis (Omdahl et al., 2002). In addition to calcium and PTH, the CYP27B1 is also negatively regulated by calcitriol itself, phosphate and fibroblast growth factor 23 (FGF23) (Brenza et al., 2000; Murayama et al., 2000). FGF23 is a phosphaturic factor that promotes renal phosphate excretion by decreasing its reabsorption in the proximal tubule. FGF23 inhibits the synthesis and promotes catabolism of calcitriol to reduce the level of 1,25(OH)2D3, which in turn decreases FGF23 expression in bone through a negative feedback circuit between the FGF23 and the vitamin D pathway (Bai et al., 2004). Although the kidney is the major source of circulating calcitriol, CYP27B1 is also expressed in several cell types, including cancer ones. Thus, the active hormone synthesized in extrarenal tissues can exert anticancer actions in an autocrine, intracrine or paracrine manner, in addition to signals that come from circulating calcitriol synthesized in the kidney. Extrarenal CYP27B1 enzyme is not regulated by the calciotropic hormones, in fact, the calcitriol synthesis mostly depends on the circulating 25(OH)D3 concentration (Adams et al., 2012; Wang et al., 2004; Young et al., 2004). The expression of CYP27B1 in cancer cells suggests that supplementation with vitamin D might be used in cancer therapy because once ingested is converted in calcidiol, increasing its serum concentration, that triggers the synthesis of the active form, 1,25(OH)2D3, within the cancer cells, where it can exert anticancer actions (Hsu et al., 2001; Whitlatch et al., 2002). Among the many genes regulated by calcitriol, CYP24A1 (also known as 24-hydroxylase) is particularly important: it protects the body from the excessive calcitriol level. CYP24A1 catalyzes the hydroxylation of both 25(OH)D3 and 1,25(OH)2D3 to form 1,24,25(OH)3D3 (also known such as calcitroic acid), a molecule with reduced biological activity and reduced affinity for the VDR and which is excreted in the bile (Pike et al., 2012; Omdahl et al., 2002). CYP24A1 is expressed in all the cells that are responsive to calcitriol and its activity is regulated, such as CYP27B1, by calcitriol itself, PTH and FGF23 (Shimada et al., 2004). The CYP24A1 enzyme is stimulated by calcitriol and inhibited by both low calcium concentration and PTH (Omdahl et al., 2002). Calcitriol, thus, inducing CYP24A1, suppresses itself through an autoregulatory feedback that attenuates gene transcriptional effects of calcitriol to protect the body against hypercalcemia. The main role of vitamin D is as a modulator of calcium homeostasis and osteo-synthesis. Without vitamin D, only 10 to 15% of dietary calcium and about 60% of phosphorus is absorbed (Holick et al.,

2006; Bouillon et al., 2001). The interaction of 1,25-dihydroxyvitamin D with the vitamin D receptor increases the efficiency of intestinal calcium absorption to 30 to 40% and phosphorus absorption to approximately 80% (Heaney et al., 2003). Serum level of Vitamin D is directly related to bone mineral density in men and women (Bischoff-Ferrari et al., 2006). When the level of calcitriol decreases, the levels of PTH increase reducing the calcium absorption from intestine and inducing the synthesis of 1,25(OH)2D3, i.e., calcitriol, in the kidney (Chapuy et al., 1997; Heaney et al., 2003; Holick et al., 2005).

Non-Canonical Metabolic Pathway

The limit of the use of Vitamin D as a therapeutic for cancer is its role in calcium homeostasis and bone mineralization. In fact, an elevated concentration of Vitamin D in the blood can be toxic and cause hypercalcemia. This has led researchers to find Vitamin D analogs that possess anti-tumor effects without significant calcemic activity. Recently an alternative pathway of Vitamin D metabolism through enzymatic activity of CYP11A1, a cytochrome P450 side chain cleavage, has been showed (Slominski et al., 2005). CYP11A1 catalyzes the hydroxylation of cholesterol to form the common precursor of steroid hormones, pregnenolone, in steroidogenesis (Miller et al., 2011). However, CYP11A1 can use Vitamin D as alternative substrate to cholesterol (Slominski et al., 2014; Tongkao-On et al., 2015). The metabolism of Vitamin D can be mediated by CYP11A1 involving sequential hydroxylation to form Vitamin D analogs. The main analogs resulting from a single hydroxylation by CYP11A1 include 20(OH)D3, 22(OH)D3 and 17(OH)D3 (Slominski et al., 2017). These are further hydroxylation by this enzyme to generate 20,22(OH)2D3, 20,23(OH)2D3, 17,20(OH)2D3. The major product of this alternative pathway is 20(OH)D, which can also serve as a substrate for another enzyme, such as CYP24A1, to be hydroxylated. Vitamin D analogues exhibit comparable or better anti-proliferative, differentiating and anti-inflammatory activity than that performed by calcitriol (Slominski et al., 2017; Slominski et al., 2017). These metabolites enhance defense mechanisms against UVB-induced DNA damage and oxidative stress and elicit anticancer properties in a cell line-dependent manner. Interestingly, 20(OH)D3 and 20,23(OH)2D3 are biased agonists of VDR, do not induce calcemic effects or the expression of CYP24A1, as in response to calcitriol (Slominski et al., 2017). Thus, the pleiotropic and different effects of vitamin D could be attributed not only to the effects of the calcitriol–VDR pathway, but also to those of CYP11A1-derived vitamin D metabolites-VDR pathways.

VITAMIN D AND CANCER: THE EPIDEMIOLOGICAL EVIDENCE

Several data strongly suggest that vitamin D deficiency increases the risk of developing multiple malignancies. Several studies demonstrated an inverse association between cancer risk and circulating levels of 25(OH)D3, reflecting both sun exposure and dietary Vitamin D intake. Epidemiological studies have suggested both increased incidence rates and elevated mortality rates in several cancers in geographical regions or in populations that are exposed to less solar ultraviolet B (UV-B) radiation. The Vitamin D-cancer correlation has been demonstrated for several cancers, including colorectal cancer, prostate cancer and breast cancer. In 1941, Apperly firstly suggested that the difference in cancer incidence observed in America and Canada at different latitudes was negatively correlated with the serum level of vitamin D in the areas with a major sun exposure analyzing population engaged in agriculture (Apperly, 1941). He demonstrated a highly significant correlation between reduced cancer mortality in adults and a career in agriculture. This analysis led him to conclude that cancer mortality could be reduced "by inducing partial or complete immunity" by exposing the skin to solar light. More recently, in the 1980's, Garland and Garland proposed a role for Vitamin D in reducing cancer risk and colon cancer mortality rates (Garland and Garland, 1980). They used an ecological approach, which compares disease outcomes averaged by geographical units with risk-modifying factors, providing perhaps the strongest evidence that solar UVB and vitamin D reduce the risk of cancer. The ecological approach is strong for cancer because solar UVB, through production of Vitamin D, is a significant factor to reduce many types of cancer, but other risk-modifying factors, such as ethnic background and smocking, can be considered in this type of analyses. Single-country ecological studies provide strong and consistent evidences for several cancers. Single countries have the advantage that the population is reasonably homogeneous, with similar diets and smoking habits. Multiple epidemiologic studies have found an inverse correlation between serum 25(OH)D3 concentration and risk of many types of cancer including breast, colorectal and prostate. While ecological studies may provide the best evidence for a role of solar UVB and vitamin D in reducing the risk of cancer, other epidemiological findings should be supported by specific investigations. The other most common type of epidemiological studies evaluates the role of solar UVB and vitamin D in reducing the cancer risk is the observational study. There are two types of observational studies: case-control studies in which the serum 25(OH)D3 level is determined near the time of cancer diagnosis; and nested case-control studies from cohort studies (Moukayed and Grant, 2013). The advantages of the nested case-control studies include unbiased matching of controls and using 25(OH)D3 levels, UVB doses or irradiances not affected by the health outcome. Case-control studies could have serum 25(OH)D3 levels affected by the disease state. However, does not appear any evidence that the onset of cancer per se affects serum 25(OH)D3 levels or behavior, since most people do not know they have cancer until it is diagnosed. The primary problem of the nested case-control study approach is that serum 25(OH)D3 levels from blood signed at time of enrollment are generally used in the analysis, no matter how many years have elapsed since enrollment. Among the epidemiological studies that have examined the vitamin D-cancer hypothesis, greatest weight has typically been bestowed on studies that measure prediagnostic circulating levels of 25(OH)D3. Among cancers, associations have most consistently been observed for colorectal cancer. Meta-analyses show a statistically robust 30–40% reduction in colorectal cancer risk in those patients with high 25(OH)D3 levels compared with those who have low 25(OH)D3 levels, even after adjustment for other known risk factors. Recent meta-analysis of 14 prospective studies of prostate cancer showed no inverse association with 25(OH)D3 levels, and some evidence even suggests an increased risk with higher 25(OH)D3 levels. A meta-analysis of 11 prospective studies evaluating vitamin D intake or vitamin D levels, conducted in the United States, Europe, and Asia, demonstrated that vitamin D intake and blood 25(OH)D3 levels are both inversely associated with risk of colorectal cancer. Eight studies comprising 3,556 patients support evidence that serum Vitamin D levels are inversely associated with risk of colorectal cancer for patients with 25(OH)D levels in the highest compared with lowest. An increase of 25(OH)D by 20ng/mL was associated with a risk reduction of 59% for rectal cancer and 22% for colon cancer. One of the largest studies, the first based on European populations, showed that compared with a mid-range concentration of 50–75nmol/L, 25(OH)D levels lower than 25ng/mL(50nmol/L) were associated with an increased risk of colorectal cancer. Patients with 25(OH)D3 levels greater than 100nmol/L had a significant 40% lower risk of colorectal cancer than those patients with levels lower than 25 nmol/L. Previous studies have also shown that low levels of vitamin D are associated with increased mortality and aggressiveness of colorectal cancer, while there was a significant 30% reduction in formation of colorectal adenomas among patients with higher versus lower 25(OH)D3 levels.

There are contradictory results for breast cancer. One meta-analysis found no appreciable association between prediagnostic circulating 25(OH)D3 levels and breast cancer risk. Furthermore, a meta-analysis of studies focusing on the relationship between breast cancer and serum levels of 25(OH)D3 – involving 1,760 subjects - showed a 50% reduced risk of developing cancer with serum 25(OH)D3 >52 ng/ml, compared to those with serum 25(OH)D <13 ng/ml. Levels of 25(OH)D3 >52 ng/ml can be achieved by oral intake of 3,000 IU/day of vitamin D analogs (equivalent to 2,000 IU/day orally in combination with exposure to sunlight for about 12 min/day). There are few data regarding the relation between vitamin D and recurrence of breast cancer after treatment. It was investigated the survival in a large prospective cohort of 1,295 postmenopausal breast cancer patients in relation to post-diagnostic serum levels of 25(OH)D3; the study showed that lower serum calcidiol concentrations were associated with poorer overall and distant disease-free survival.

Some emerging findings suggest that higher 25(OH)D3 concentrations before or at the time of diagnosis and treatment might improve survival from various cancers. In one study of patients with prostate cancer, men had lower of prediagnostic 25(OH)D3 levels had a 60% higher risk of cancer death than those in the highest quartile. In patients with colorectal cancer, in three large cohorts, patients with higher prediagnostic 25(OH)D3 levels (>30 ng per mL) had a 30–40% lower risk of cancer-specific and total mortality.

Preventive Studies: Randomized Clinical Trials and Vitamin D

However, in order to confirm the correlation between vitamin D and the development of cancer risk, studies from randomized controlled trials (RCTs) with vitamin D supplementation are mandatory. Several clinical trials were performed to analyze the effects of vitamin D3 or of calcitriol alone or in combination with other nutrients on diverse malignancies. Unfortunately, RCTs on vitamin D, in general, do not support observational studies (Autier et al., 2014). The WHI (Women's Health Initiative) clinical trial, conducted on 36,282 postmenopausal women for 7 years, showed no significant protective effect of supplementation with vitamin D (400 IU/day) on the development of colorectal and breast cancer (Brunner et al. 2011). This study had many issues: (1) a small portion of enrolled women responded to treatment (2) the simultaneous intake of estrogen prejudiced the effects of treatment (3) women are treated with a too low dose of vitamin D (400 IU/day) to raise the concentration of vitamin D, correcting the deficit, and to promote an anticancer effect. The first vitamin D RCTs reported significant reductions in cancer incidence was a 4-year trial involving postmenopausal women in Nebraska (Lappe et al, 2007). The trial had three groups: 1000 IU/d of vitamin D3 plus 1450 mg/d of calcium, 1450 mg/d of calcium, and placebo, [mean baseline 25(OH)D concentration = 29 ng/mL]. Serum 25(OH)D concentrations were measured at baseline and annually. The researchers analyzed results for all 4 years, and for the last 3 years. In the other vitamin D RCT of postmenopausal women in Nebraska, participants took 2000 IU/d of vitamin D + 1500 mg/day of calcium in the treatment group or placebo. The mean baseline 25(OH)D concentration was 33±10ng/mL and rose to 44 ng/mL in the treatment group. Cancer incidence was lower in the Vitamin D + calcium women group than in the placebo control subjects. In postmenopausal women, thus, the supplementation with calcium and vitamin D substantially reduces cancer risk (Lappe et al., 2017). In a small clinical trial, prostate cancer patients are treated with Vitamin D3 at dose 2000 IU/day for a period of 21 months (Woo et al. 2005). The Vitamin D3 supplementation has beneficial effects to decrease the prostate specific antigen (PSA) levels in the patients treated. Serum PSA is a biomarker of prostate cell abundance, and its rate of rise is often used as a surrogate for tumor growth.

Another study analyzed vitamin D3 supplementation for advanced or metastatic colorectal cancer patients. In this trial were enrolled both men and women. Patients randomized to high-dose vitamin D (4,000 IU/d) supplementation had increased 25(OH)D3 level and improved progression-free survival in comparison those received low-dose Vitamin D (400 IU/d) supplementation.

The VITamin D and OmegA-3 TriaL (VITAL) (25871 participants) is a RCT of Vitamin D3 at a dose of 2000 IU /day and omega-3 fatty acid at a dose of 1g/day supplements in the prevention of cancer among 50 years of age or older men and women in the USA. VITAL tests the independent effects of vitamin D and omega-3 fatty acid supplementation on risk for developing cancer (Manson et al., 2012) This trial also examines whether vitamin D and omega-3 fatty acid supplements exhibit synergistic or additive effects on cancer and whether the effect of each supplement on cancer risk varies by baseline blood levels of vitamin D and eicosapentaenoic acid plus docosahexaenoic acid (EPA+DHA).

Supplementation with vitamin D was not associated with a lower developing cancer risk than placebo supplementation.

It has been pointed out that the improper design of RCTs accounts to their predictable failure in confirming the results of observational studies. Those studies were indeed designed as standard RCT models for pharmaceuticals rather than on guidelines appropriate for nutrients, e.g., the drug model assumes that the only source of the agent is in the trial and that a linear dose-response relationship exists (Heaney, 2014). However, vitamin D trials can satisfy neither assumption. In addition, it is not vitamin D dose, per se, that affects health outcomes but rather vitamin D status over time [serum concentrations of 25(OH)D] and, to a lesser extent, serum 1,25-dihydroxyvitamin D. Heaney's guidelines, as applied to vitamin D, include:

- Ensure an understanding of the 25(OH)D concentration ± health outcome relation of interest.
- Measure 25(OH)D concentrations of prospective participants.
- Aim to include only those with 25(OH)D concentrations near the low end of the relationship.
- Supplement those in the treatment arm with enough vitamin D3 to increase 25(OH)D concentrations to the upper region of the relationship.
- Measure achieved 25(OH)D concentrations.
- Optimize co-nutrient status (such as calcium).

Anticancer Effects of Vitamin D

In vitro studies demonstrated that exposure of tumor cells to high vitamin D concentration inhibits their proliferation and induces differentiation. In 1981, Colston and collaborators, in fact, demonstrated that the addition of calcitriol to mouse myeloid leukemia cells inhibited their proliferation (Colston et al., 1981). Genomic and proteomic screening approaches have identified an extensive array of VDR target genes that mediate the antineoplastic action of calcitriol. Vitamin D can regulate several cellular mechanisms such as proliferation, differentiation, apoptosis and modulate cell-microenvironment interaction (Giammanco et al., 2015).

The cell cycle is regulated by a complex network of interlinked regulators that in concert mediate cellular proliferation. The major mechanism through which Vitamin D exerts its anticancer action includes the blocking of cancer cells proliferation (Samuel et al., 2008). This function is mediated by the regulation of growth factors, cell cycle and signaling pathways. Vitamin D, for example, inhibits the mitogenic signaling by growth factors such as Insulin-like Growth Factor (IGF) (Peng et al., 2004) and increases the cyclin-dependent kinase (CDK) inhibitors, such as p21 and p27 (Liu et al., 1996), leading to the dephosphorylation of the retinoblastoma protein (Rb) and G0/G1 phase cell cycle arrest (Krishan et al., 2007). Vitamin D also up regulated the growth-arrest DNA damage repair factor, GADD45α, at both the mRNA and protein levels.

FoxO proteins are tumor-suppressors that control cell proliferation. The function of several FoxO family members is inhibited by MAPK-mediated phosphorylation. 1,25(OH)2D3 treatment regulates binding of FoxO3a and FoxO4 to DNA regulatory regions by stimulating a direct interaction between VDR, FoxO3a or FoxO4, and the FoxO regulators Sirt1 (a class III histone deacetylase) and protein phosphatase 1. Sirt1 and protein phosphatase 1 promote nuclear retention of FoxO proteins by counteracting MAPK-mediated phosphorylation. It was also demonstrated to inhibit telomerase activity by reducing the expression of telomerase reverse transcriptase (TERT) and to induce the expression of transforming growth factor β (TGFβ) and its receptors, leading to inhibition of cell growth (Welsh, 2012).

In response to calcitriol, some cancer cells acquire a less malignant and more normal phenotype, inducing a pro-differentiating effect. Vitamin D induces the differentiation by the regulation of different intracellular signaling pathways, such as phosphatidylinositol3 kinase/AKT (PI3K/AKT), MAPK (ERK), NF-kB, and Ca2+signaling (O'Kelly et al., 2005). Studies on VDR-null mice show a role of calcitriol in physiological apoptosis in normal mammary gland development. Vitamin D triggers apoptosis in many but not all cancer cells by cell type-specific mechanisms, including the activation of intrinsic pathway of apoptosis through suppression of anti-apoptotic genes such as BCL2 and the stimulation of the pro-apoptotic gene such as BAX, BAD and BAK (Blutt et al., 2000). Vitamin D can also inhibit the AKT-mediated anti-apoptotic signaling pathway by increasing the expression of phosphatase and tensin homolog (PTEN), a tumor suppressor gene. Moreover, Vitamin D can initiate apoptotic events by recruiting Ca2+-dependent apoptotic effectors such as calpain and caspase 12. An alternative mechanism through vitamin D mediates cell growth arrest is the disruption of β-catenin function, the terminal mediator of Wnt signaling. In the cytoplasm, β-catenin is found in association with adenomatous polyposis coli (APC). Activation of Wnt signaling leads to the accumulation of β-catenin and its release from APC. This free β-catenin translocate to the nucleus, binds with the transcription factor TCF4 on DNA, and activates transcription of genes whose protein products control proliferation (e.g., c-myc and cyclin D1). Calcitriol inhibits the Wnt/β-catenin signaling pathway affecting the expression of known regulators of Wnt-signaling such as the Wnt activator dickkopf-4 (DKK-4) and increases the expression of Wnt antagonist dickkopf-1 (DKK-1). Wnt promotes cell malignancy and metastasis triggering matrix metallo-proteinase 7 (MMP7). 1,25(OH)2D3-mediated events may also indirectly influence β-catenin function through increased production of E-cadherin, a membrane protein that can bind β-catenin and prevent its nuclear accumulation and required for cellular adhesion, modulating the metastatic transformation associated with Wnt-activated tumor cells. Vitamin D, counteracting Wnt signaling can decrease proliferation and can inhibit invasion and metastasis process of cancer cells. The antitumor effects of Vitamin D that antagonize Wnt signaling action have been demonstrated in breast, colon and prostate cancers. Thus, Vitamin D treatment downregulates several of the above tumorigenesis-promoting genes.

Inflammation contributes to the development and progression of many cancer types (Mantovani et al., 2008). Calcitriol shows beneficial anti-inflammatory actions in cancers via several mechanisms. Calcitriol, for example, inhibits the prostaglandin (PG) pathway involved in pro-inflammatory responses. PGs promote carcinogenesis and play a positive role in the progression of many cancers by stimulating cellular proliferation, promoting angiogenesis and inhibiting apoptosis. Cyclooxygenase, the enzyme responsible in PG synthesis, is an oncogene and an important molecular target in cancer therapy. COX-2 is the rate-limiting enzyme of PGs synthesis and an immediate-early response gene whose expression is rapidly induced by mitogens, cytokines and growth factors. Calcitriol inhibits the synthesis and biological actions of pro-inflammatory PGs (Krishnan et al., 2010). Calcitriol, for example, suppresses COX-2 expression, upregulates the expression of 15-hydroxyprostaglandin dehydrogenase (15-PGDH), the enzyme that inactivates PGs and enhances the expression of PG receptors that are essential for PG signaling (Krishnan et al., 2010; Moreno et al., 2005). These calcitriol actions contribute to the suppression of the proliferative and angiogenic stimuli provided by PGs in malignant cells and constitute important pathways that mediate the anti-inflammatory effects of calcitriol. Moreover, Vitamin D can regulate the p38-MAPK-mediated pro-inflammatory signaling pathway. In both normal and tumoral prostate epithelial cells, vitamin D increases the expression of MAP kinase phosphatase 5 (MKP5), which dephosphorylates and thereby inactivates MAPKs (Peehl et al., 2004). MKP5 specifically dephosphorylates p38 MAPK and the stress-activated protein kinase Jun-N-terminal kinase (JNK), leading to their inactivation (Nonn et al., 2006). Calcitriol increases MKP5 transcription by binding to and activating the VDR and its subsequent interaction with a vitamin D response element (VDRE) identified in the MKP5 promoter (Nonn et al., 2006). These functions lead to downstream anti-inflammatory responses by causing the dephosphorylation and inactivation of the p38 stress-induced kinase, resulting in a decrease in the production of pro-inflammatory cytokines that sustain and amplify the inflammatory response, such as interleukin-6 (IL-6) (Parl et al., 2003). IL-6 is a major pro-inflammatory cytokine that participates in inflammation-associated carcinogenesis, and it has been implicated in the pathogenesis of several cancers (Rose-John et al., 2007). Serum IL-6 levels were significantly elevated and positively correlated to tumor burden in Colorectal, Breast and Prostate cancer patients, who also exhibited a positive correlation between IL-6 levels and the number of bone metastases. Calcitriol can also inhibit the nuclear factor kappa B (NF-kB) signaling pathway through several mechanisms (Karim and Lin, 2002). NF-κB comprises a family of inducible transcription factors ubiquitously present in all cells (Wang et al., 2014). NF-κB transcription factors are important regulators of innate immune responses and inflammation. In normal cells, most NF-κB dimers are bound to specific inhibitory proteins named IκB proteins, and pro-inflammatory signals activate NF-κB mainly through IκB kinase (IKK)-dependent phosphorylation and degradation of the inhibitory IκB proteins. Free NF-κB then translocate to the nucleus and activates the transcription of pro-inflammatory cytokines, chemokines, and anti-apoptotic factors. In contrast to normal cells, many malignant cells have elevated levels of active NF-κB (Palayoor et al., 1999). Calcitriol blocks NF-kB activation by increasing the expression of IkBa. In fibroblasts, however, Vitamin D induces the binding of VDR to IkBa Kinase (IKK), preventing its phosphorylation and activation and thereby inhibiting the nuclear translocation of the p65 subunit of NF-kB, thereby inhibiting the NF-κB complex from binding to its DNA response element and consequently suppressing the NF-κB stimulation of transcription of downstream targets such as IL-8. Moreover, Vitamin D can regulate the interaction between immune and cancer cells to suppress the production of pro-inflammatory cytokines. Co-culture experiments using immune cells and colon cancer cells, revealed that vitamin D treatment decreased the production of cytokines, supporting the role of anti-inflammation of Vitamin D in tumor microenvironment.

Hypoxia and oxidative stress are often associated with cancer progression. Thus, maintaining antioxidant defense systems should be a critical step in preventing tumor development. Oxidative stress often induces DNA damage and loss of DNA-repair ability. Accumulating data suggest the vitamin D can protect from oxidative stress-induce DNA damage by promoting antioxidant defenses. Treatment of several cancer cell lines with vitamin D activates cellular signaling cascades to induce the expression of numerous enzymes involved in reactive oxygen species (ROS) detoxification. Vitamin D reduces thioredoxin and promotes antioxidant responses, induces mRNA expression of superoxide dismutase (in prostate epithelial cells) and down-regulates glutathione levels by increasing glucose-6-phosphate dehydrogenase (G6PD) expression. Moreover, Vitamin D increases its expression, nuclear translocation, and decreases the degradation of NF-E2-related factor (NRF2), the master regulator of the expression of antioxidant enzymes. This can be a potential mechanism by which Vitamin D can induce antioxidant enzymes, thus exerting its antioxidant action.

Angiogenesis is essential for cancer cell growth and metastasis (Folkman, 1995). Angiogenesis is a multistep, multicellular process that depends upon a variety of pro-angiogenic factors, including vascular endothelial growth factor (VEGF), basic fibroblast growth factor (bFGF), and platelet derived growth factor BB homodimer (PDGF BB). Increased expression of pro-angiogenic factors such as HIF1α, VEGF, angiopoietin 1 and platelet-derived growth factor (PDGF) in tumors and in VDR-null mice suggest regulation of these molecules by calcitriol–VDR complex. 1,25(OH)2D3 inhibit the development of the tumor vasculature required for the progression of solid. Vitamin D, in fact, inhibits initiators of cellular angiogenesis in several cancer cell lines suppressing, for example, the expression of VEGF (Mantell et al., 2000). Calcitriol can also act through an indirect mechanism by decreasing prostaglandin E (PGE) generated by COX-2 which triggers angiogenesis by increasing HIF1a synthesis in cancer cells (Sahin et al., 2009). MMPs promote angiogenesis by mediating the degradation of the basement membrane of the vascular epithelium and the extracellular matrix, thereby creating a passageway in these barriers for the formation of new capillaries. In cancer, calcitriol decreases the expression and activity of MMP-9 while increases the activity of its counterpart tissue inhibitor of metalloproteinase 1 (TIMP-1), thereby decreasing angiogenesis and the invasive potential of these cells (Sakamoto et al., 2008).

In the last years, it was demonstrated that calcitriol also exert its anticancer effect by regulating microRNAs (miRNAs) functions; miRNAs are short single strands of RNA that function as regulators of mRNA translation. They have several roles in many biological processes (Yates et al., 2013). Calcitriol can increase or decrease the expression of miRNAs implicated in Vitamin D pathway (Maier et al., 2009). For example, in colon cancer cells miR-22 and miR-627 expression is enhanced by calcitriol, positively correlated with VDR levels and these miRNAs are implicated, respectively in the inhibition of cancer cells migration and in the anti-proliferative action both mediated by calcitriol (Alvarez-Diaz et al., 2012; Padi et al., 2013). Interestingly, miRNAs have also been implicated in the regulation of VDR and CYP24A1 expression, suggesting an impact of miRNAs on the efficacy of vitamin D therapy. Post-transcriptional regulation by miRNAs is associated with the CYP24A1 overexpression in cancer cells. MiR-125b, frequently downregulated in many cancers, binds to the 3'-UTR of CYP24A1 mRNA and thereby inhibit its translation (Komagata et al., 2009). Indeed, CYP24A1 expression levels are inversely correlated to miR-125b levels in breast cancer tissues, thus, low levels of miR-125b promote CYP24A1 overexpression in cancer. miR-92 is another miRNA capable of regulating CYP24A1 overexpression (Borkowski et al., 2015). Thus, miR-92 inhibitors decrease CYP24A1 expression levels. It will be of great interest to determine whether the expression of any miRNAs is correlated with clinical outcomes and whether these miRNAs can inform the ongoing clinical trials to make better predictions about which patients are likely to respond to vitamin D and calcitriol therapy.

Dysregulation of Vitamin D Pathways in Cancer

The enzymes of Vitamin D metabolism are widely expressed in many tissues other than the kidney and its target tissues such as bone and intestine. This suggests that most cells of the human organism could be target of Vitamin D and that calcitriol levels could be regulated to exert tissue-specific functions. The local regulation of Vitamin D activity is dysregulated in many cancer cells, contributing to the resistance to cancer therapy based on Vitamin D.

VDR is expressed in most cells, but its expression is progressively reduced during tumor progression in many cancer types. Studies on ovarian, prostate, breast and skin cancer, revealed a negative correlation between

VDR expression and tumor malignancy (Lopes et al., 2010; Hendrickson et al., 2011). High VDR expression is associated with reduced risk of lethal prostate cancer progression and poor diagnosis (Srinivasan et al., 2011). Cancer cells may counteract the tumor-suppressive effects of vitamin D by developing several mechanisms to reduce VDR expression. Many different factors have been proposed to influence VDR expression in cancer. First, over-expression of SNAIL transcription factors can reduce VDR gene expression and blunt 1,25(OH)2D3 or vitamin D analog-induced cell growth arrest and/or gene expression of cancer cells. Snail2, for example, involved in epithelial-mesenchymal transition (EMT) metastasis and tumor invasion, inhibits VDR expression in cancer (Palmer et al., 2004). Snail can bind to the promoter region of VDR gene to recruit co-repressors inhibiting its transcription in colon and breast cancer. Another protein that regulates VDR gene expression is p53 (Maruyama et al., 2006). The p53 protein protects the cells against DNA damage inducing apoptosis. Tumor-suppressor p53 is lost or mutated in almost half of all tumors because of increased degradation or the presence of inactivating mutation on its gene (Stambolsky et al., 2010). A mutated p53 can cause deregulation on the anticancer activity of the VDR pathway increasing nuclear accumulation of VDR, which in some cases correlates with tumor stage. In order to inhibit apoptosis, high endogenous levels of mutated p53 in cancer cells probably cooperate with Vitamin D, enhanced by supplementation. Epigenetic regulation plays also a role in VDR expression in cancer. CpG Island methylation in the VDR promoter region is associated with its reduced expression in colon and breast cancer cells (Liel et al., 1999). Inhibitors of DNMT induce VDR expression and enhance the anti-proliferative effect of calcitriol in breast cancer cells (Marik et al., 2010).

All the cells that express CYP27B1 may increase the local concentration of 1,25(OH)2D3. The expression of CYP27B1 is inversely correlated with the progression of many types of cancer, suggesting an important role of Vitamin D on cancer prevention (Hsu et al., 2001). Recent studies demonstrated the downregulation of CYP27B1 in colon cancer cells by pro-inflammatory cytokines, thus suggesting how microenvironment could contribute in decreasing enzyme levels during tumor progression (Hummel et al., 2014). CYP27B1 enzymatic activity in cancer depends on the organ and the tumor grade, in fact, the expression level can be decreased, increased or unchanged. In well-differentiated tumors, the CYP27B1 expression level is higher than highly aggressive and poorly differentiated tumors (Cross et al., 2001).

CYP24A1 or 24-hydroxylase, the catalyzing enzyme of calcitriol and calcidiol, presents an aberrantly high basal expression level in several cancer cells (Miller et al., 1995; Skowronski et al., 1993). Cancer cells may upregulate CYP24A1 expression to reduce local concentration of 1,25(OH)2D3 and to obtain resistance to it, like the reduction of CYP27B1 in some cancers (Friedrich et al., 2003). CYP24A1 is amplified in breast cancer and it can be proposed as an oncogene (Albertson et al., 2000). Upregulation of CYP24A1 in some cancers is correlated with poor clinical outcome. These findings suggest that cancer cells can induce the expression of CYP24A1, reducing the local concentration of vitamin D, to bypass its anticancer effect. However, co-addition of inhibitors of 24-hydroxylase — such as liarozole, ketoconazole — makes the cells more responsive to calcitriol (Ly et al., 1999; Swami et al., 2005). These observations suggest that combinations of calcitriol with inhibitors of 24-hydroxylase may be a useful strategy in cancer treatment (Horvath et al., 2010). The combination therapy may also allow the use of calcitriol at lower concentrations and thereby reduce its hypercalcemic side effects (Wang et al., 2012). However, the inhibition of CYP24A1 enzymatic activity amplifies the biological activity of calcitriol but the use of cytochrome P450 inhibitors can cause calcitriol-resistant cells and enhance its calcemic effect not only its anticancer actions (Muindi et al., 2010). In normal cells CYP24A1 expression is induce by calcitriol-VDR activation, but in cancer cells the high level of this enzyme is unlikely to be mediated by VDR activation since that VDR expression and activity is downregulated in most type of cancers. This suggests that its over-expression cannot be the result of normal physiological processes. Several mechanisms can be responsible for the induction of CYP24A1 in cancer. Epigenetic regulation involving the CYP24A1 promoter region contributes to its altered expression in cancers (Luo et al., 2010). Data suggested that the expression of CYP24A1 is inversely correlated with the methylation of its promoter region in lung and prostate cancer. For example, inhibition of DNA methyltransferase (DNMT) increases the expression of CYP24A1 in lung and colon cancer (Hobaus et al., 2013).

VITAMIN D AND ANIMAL MODELS OF CANCER

The effect of calcitriol and its analogues on the inhibition of tumor growth has been extensively studied in animal models to evaluate the anticancer effects in vivo. Data from animal models of several cancers reveal that Vitamin D, calcitriol and its analogues cause a significant reduction in tumor growth. For example, the VDR-null mice show hyperproliferation in mammary glands, the parathyroid gland and the descending colon. VDR-/- mice displayed colon cell division and increased expression of cell proliferation markers, such as proliferating cell nuclear antigen (PCNA) and Cyclin D. Moreover, deletion of the VDR gene predisposes mice to higher risk and susceptibility to chemically induced cancers of the skin, prostate, blood and breast. For example, treatment of VDR-knockout mice with chemical carcinogen, leads to dysregulated skin proliferation and hyper-proliferation of both basal and epidermal cells. However, VDR-/- mice do not spontaneously develop skin tumors without carcinogens predisposition. In mice with pRB and p53 tumor-suppressor inactivated, prostate cancer has a cellular proliferation rate and tumor progression higher and faster than in wild-type mice. Treatment with Vitamin D in combination with other agents, such as calcium, seem to exert significantly greater beneficial anticancer effects than use of Vitamin D alone. Supplementation of Vitamin D with calcium, in fact, inhibits tumor development and growth. This finding supports the hypothesis that Vitamin D and calcitriol could inhibit cancer development and progression.

Calcitriol Effect on Breast, Prostate, and Colon Cancer

Calcitriol has a wide range of actions in cancer cells from many different types of cancer. Calcitriol regulates specific signalling pathway in breast, prostate and colon cancer.

Breast cancer is one of the most common female cancers in the Western population. Several studies have shown that 1,25(OH)2D3 induces differentiation and apoptosis and inhibits cell proliferation and angiogenesis in normal and malignant breast cells through vitamin D receptor (VDR). In both normal and breast cancer the cells express CYP27B1, the enzyme that form 1,25(OH)2D3, which regulates development of mammary gland. Several data have demonstrated an increased expression of CYP24A1 in cancer cells compared to healthy cells, suggesting that in malignant tissue Vitamin D is actively degraded. Calcitriol, in breast cancer, inhibits both the synthesis and the biological function of estrogens, the major stimulators of growth of this cancer. Calcitriol, in fact, suppresses the estrogen pathway through the repression of expression of the gene encoding aromatase (CYP19A1), the cytochrome P450 enzyme that catalyzes estrogen synthesis from androgenic precursors. Calcitriol also downregulates ERα, the nuclear receptor that mediates estrogen effects. Aromatase expression is higher in breast cancer cells than in normal tissue. Moreover, aromatase is critical for the progression of ER+ breast cancer. Calcitriol significantly decreases aromatase expression in both ER+ and ER-negative human breast cancer cells. In breast cancer cells calcitriol down-regulates aromatase through a direct repression of aromatase transcription via promoter II through the vitamin D response elements (VDREs) present in this promoter. Anti-inflammatory effects of calcitriol are shown in breast cancer cells. In both ER+ and ER negative human breast cancer cells, calcitriol down-regulates the expression of cyclooxygenase-2 (COX-2), the rate-limiting enzyme catalyzing PG synthesis. Calcitriol also increases the expression of 15-PGDH, which catalyzes the conversion of PGs to biologically inactive keto-derivatives. As a result, calcitriol treatment significantly reduces both the synthesis and the biological activity of PGs in breast cancer cells. PGs play an important role in the development and progression of breast cancer. PGs released from breast cancer cells or surrounding tissues mediate autocrine/paracrine stimulation of tumor progression by promoting cell proliferation and resistance to apoptosis and stimulating tumor cell migration, metastasis and angiogenesis. Elevated expression of COX-2 in breast cancer is associated with larger tumor size and poor clinical outcome. COX-2 may be an important factor in promoting tumor progression in ER-negative tumors and is a potential drug target in breast cancer therapy. The growth stimulating actions of estrogen require also the action of ERα, the specific nuclear receptor that mediates proliferation in response to estrogens. Calcitriol down-regulates ERα expression in breast cancer cells. Calcitriol and its metabolites suppress estrogenic actions in breast cancer cells inhibiting the expression of estrogen-responsive genes such as the progesterone receptor and attenuating the stimulation of the growth of breast cancer cells by estradiol (E2). Calcitriol down-regulate ER through a direct transcriptional repression of the ERα gene. Studies have identified negative vitamin D response elements (nVDREs) in the ERα promoter that mediate the repressive effects of calcitriol on the ERα gene. Calcitriol causes a decrease both in the level of locally produced hormone (estrogens) by the breast cancer epithelial cells and the surrounding breast adipose tissue and in the levels of receptors through which they act (ERα) in breast cancer cells. Thus, the downregulation of both the hormone and the receptor levels significantly reduces the important proliferative stimulus of estrogens on ER+ breast cancer cells.

In Western countries, prostate cancer is the most common malignant tumor in men and a major cause of cancer deaths. Studies concerning the role of Vitamin D in prostate cancer have rapidly increased because the deficiency of Vitamin D exacerbates the risk of this type of cancer. Both healthy and prostate cancer cells have receptors for Vitamin D and express key enzyme for Vitamin D metabolism. In prostate cancer patients, serum 25(OH)D3 concentration is usually lower than the control ones. In addition, serum IL-8 level is elevated in prostate cancer patients. This finding provides evidence that low Vitamin D status is associated with inflammation in prostate cancer. Chronic inflammation promotes progression of cancer. IL-8 promotes androgen-independent proliferation of prostate cancer cells via inducing androgen receptor expression and activation. Androgens drive the growth of prostate cancer cells through Androgen Receptor (AR)-mediated pathways. Androgen receptor (AR) deregulation and aberrant androgen synthesis significantly contribute to the progression of prostate cancer. There is a crosstalk between calcitriol and androgen signaling in prostate cancer cells. Calcitriol upregulates expression of AR and other androgen-responsive genes and, in turn, VDR is regulated by androgens. Calcitriol induces an increase in androgen-stimulated prostate-specific antigen (PSA) expression and promotes the differentiation of prostate cell progenitor into AR+ luminal epithelial cells.

Several data demonstrated a protective role of Vitamin D in colorectal cancer. Interesting is the observation that mortality for colorectal cancer increased with progressive decrease of Vitamin D levels. Vitamin D and its metabolites act as inhibitors of colorectal cancer progression via several mechanism. Calcitriol, for example, has multiple pro-differentiation effects in colorectal cancer cells. One of the pathways disrupted in colon cancer is the Wnt/β-catenin signaling pathway. In colon cancer, the Wnt/β-catenin pathway is disrupted due to mutations in β-catenin or APC. These mutations prevent the phosphorylation of β-catenin and contribute to its accumulation in the cell's cytosol; not phosphorylated β-catenin is then able to translocate and be accumulated in the nucleus. In the nucleus, β-catenin dimerizes with DNA-bound T-cell factor (TCF), triggering the expression of target genes (eg, c-myc, cyclin D1) and there it is capable to induce the transformation of normal cells into an oncogenic type. Recent research suggest that the vitamin D can modulate Wnt/β-catenin signaling. Calcitriol inhibits β-catenin pathway, thus VDR binds to β-catenin inhibiting β-catenin nuclear translocation and it increases levels of extracellular Wnt inhibitors DICKKOPF1 and DICKKOPF4. Moreover, calcitriol regulates RHOA-ROCK-p38 MAPK-MSK pathway to increase E-cadherin and other adhesions proteins to hold β-catenin on the membrane. Vitamin D in colon cancer cells inhibits the proliferation increasing, for example, the expression of the cyclin-dependent kinase (CDK) inhibitors CDKN1B and CDKN1A, thus leading to G1 phase arrest. Calcitriol also activates the latent transforming growth factor b1, (TGFb1) and sensitizes these cells to the growth inhibitory effect of this molecule. Calcitriol induces apoptosis in colorectal adenoma and colorectal cancer by upregulating the pro-apoptotic proteins, such as BAK and BAX, and by downregulating the anti-apoptotic proteins, such as BCL2.

Conclusion

Many studies indicate the importance of Vitamin D pathway in several diseases, such as cancer. Extensive research provides evidence for the anti-proliferative, anti-inflammatory, and pro-differentiation effects of calcitriol, supporting its potential role in cancer prevention and treatment. The limit in the use of Vitamin D is its effect on hypercalcemia. However, due to its potential benefits, one can argue that we should investigate alternative pathways or analogs, in order to "dissociate" calcium homeostatic activity from the anticancer effects.

References

Adams, John S, Clemens TL, Parrish JA, Holick MF. "Vitamin D synthesis and metabolism after ultraviolet irradiation of normal and Vitamin-D-deficient subjects." *N Engl J Med* 306:722–725 (1982).

Adams, John S and Hewison M. "Extrarenal expression of the 25-hydroxyvitamin D-1-hydroxylase." *Arch. Biochem. Biophys*. 523, 95–102 (2012).

Albertson, Donna G, Ylstra B, Segraves R, Collins C, Dairkee SH, Kowbel D, Kuo WL, Gray JW, Pinkel D. "Quantitative mapping of amplicon structure by array CGH identifies CYP24 as a candidate oncogene." *Nature Genet*. 25, 144–146 (2000).

Alvarez-Diaz, Silvia, Valle N, Ferrer-Mayorga G, Lombardía L, Herrera M, Domínguez O, Segura MF, Bonilla F, Hernando E, Muñoz A. "MicroRNA-22 is induced by vitamin D and contributes to its antiproliferative, antimigratory and gene regulatory effects in colon cancer cells." *Hum. Mol. Genet.* 21, 2157–2165 (2012).

Apperly, Frank L. "The relation of solar radiation to cancer mortality in North American." *Cancer Res.* 1, 191–195 (1941).

Autier, Philippe, Boniol M, Pizot C, Mullie P. "Vitamin D status and ill health: a systematic review." *The Lancet Diabetes & endocrinology* 2(1):76±89 (2014).

Bai, Xiuying, Miao D, Li J, Goltzman D, Karaplis AC. "Transgenic mice overexpressing human fibroblast growth factor 23 (R176Q) delineate a putative role for parathyroid hormone in renal phosphate wasting disorders." *Endocrinology* 145:5269–79 (2004).

Bischoff-Ferrari, Heike A, Giovannucci E, Willett WC, Dietrich T, Dawson-Hughes B. "Estimation of optimal serum concentrations of 25-hydroxyvitamin D for multiple health outcomes." *Am J Clin Nutr* 84:18-28 (2006).

Blutt, Sarah E, McDonnell TJ, Polek TC, and Weigel NL. "Calcitriol-induced apoptosis in LNCaP cells is blocked by overexpression of Bcl-2." *Endocrinology* 141, 10–17 (2000).

Borkowski, Robert, Du L, Zhao Z, McMillan E1, Kosti A, Yang CR, Suraokar M, Wistuba II, Gazdar AF, Minna JD, White MA, Pertsemlidis A. "Genetic mutation of p53 and suppression of the miR-17 approximately 92 cluster are synthetic lethal in non-small cell lung cancer due to upregulation of vitamin D Signaling." *Cancer Res.* 75, 666–675 (2015).

Bouillon, Roger. "Vitamin D: from photosynthesis, metabolism, and action to clinical applications." In: DeGroot LJ, Jameson JL, eds. *Endocrinology*. Philadelphia: W.B. Saunders 1009-28 (2001).

Bouillon, Roger, Carmeliet G, Verlinden L, van Etten E, Verstuyf A, Luderer HF, Lieben L, Mathieu C, Demay M. "Vitamin D and human health: lessons from vitamin D receptor null mice." *Endocr. Rev.* 29, 726–776 (2008).

Boyle, Iain T, Gray RW, DeLuca HF. "Regulation by calcium of in vivo synthesis of 1,25-dihydroxycholecalciferol and 24,25-dihydroxycholecalciferol." *Proc Natl Acad Sci USA* 68:2131–4. (1971).

Brenza, Holly L, DeLuca HF. "Regulation of 25-hydroxyvitamin D3 1alpha-hydroxylase gene expression by parathyroid hormone and 1,25-dihydroxyvitamin D3." *Arch Biochem Biophys* 381:143–52 (2000).

Brunner, Robert L, Wactawski-Wende J, Caan BJ, Cochrane BB, Chlebowski RT, Gass ML, Jacobs ET, LaCroix AZ, Lane D, Larson J, Margolis KL, Millen AE, Sarto GE, Vitolins MZ, Wallace RB "The effect of calcium plus vitamin D on risk for invasive cancer: results of the Women's Health Initiative (WHI) calcium plus vitamin D randomized clinical trial." *Nutr. Cancer* 63, 827–841 (2011).

Carlberg, Carsten, Bendik I, Wyss A, Meier E, Sturzenbecker LJ, Grippo JF, Hunziker W. "Two nuclear signalling pathways for vitamin D." *Nature* 361:657–660 (1993).

Christakos, Sylvia, Dhawan P, Liu Y, Peng X, Porta A. "New insights into the mechanisms of vitamin D action." *J. Cell. Biochem* 88, 695–705 (2003).

Christakos, Sylvia, Dhawan P, Verstuyf A, Verlinden L, Carmeliet G. "Vitamin D: Metabolism, Molecular Mechanism of Action, and Pleiotropic Effects." *Physiol Rev* 96:365–408 (2016).

Colston, Kay, Colston MJ & Feldman D. "1,25-dihydroxyvitamin D3 and malignant melanoma: the presence of receptors and inhibition of cell growth in culture." *Endocrinology* 108, 1083–1086 (1981).

Cross, Heide S, Bareis P, Hofer H, Bischof MG, Bajna E, Kriwanek S, Bonner E, Peterlik M. "25-Hydroxyvitamin D3-1α-hydroxylase and vitamin D receptor gene expression in human colonic mucosa is elevated during early cancerogenesis." *Steroids* 66, 287–292 (2001).

Deeb, Kristin K, Trump DL, Johnson CS. "Vitamin D signalling pathways in cancer: potential for anticancer therapeutics." *Nature reviews Cancer* 7:684–700 (2007).

DeLuca, Hector F. "Overview of general physiologic features and functions of vitamin D." *Am. J. Clin. Nutr* 80, 168
9S–1696S (2004).

DeLuca, Hector F. "Evolution of our understanding of vitamin D." *Nutr Rev* 66: S73 87 (2008).

Dusso, Adrian, Gonzalez EA and Martin KJ. "Vitamin D in chronic kidney disease." *Best Pract. Res. Clin. Endocrinol. Metab*. 25, 647–655 (2011).

Folkman, Judah. "Angiogenesis in cancer, vascular, rheumatoid and other disease." *Nat. Med*. 1:27-31 (1995).

Fraser, David R, Kodicek E. "Unique biosynthesis by kidney of a biological active vitamin D metabolite." *Nature*.; 228:764–766 (1970).

Friedrich, Micheal, Rafi L, Mitschele T, Tilgen W, Schmidt W, Reichrath J "Analysis of the vitamin D system in cervical carcinomas, breast cancer and ovarian cancer." *Recent Results Cancer Res*. 164, 239–246 (2003).

Garland, Cedric F. and Garland FC. "Do sunlight and vitamin D reduce the likelihood of colon cancer?" *Int. J. Epidemiol*. 9, 227–231 (1980).

Giammanco, Marco, Di Majo D, La Guardia M, Aiello S, Crescimannno M, Flandina C, Tumminello FM, Leto G. "Vitamin D in cancer chemoprevention." *Pharm. Biol*. 53, 1399–1434 (2015).

Glerup, Henning, Mikkelsen K, Poulsen L, Hass E, Overbeck S, Thomsen J, Charles P, Eriksen EF "Commonly recommended daily intake of vitamin D is not sufficient if sunlight exposure is limited." *J. Intern Med* 247:260-8 (2000).

Haussler, Mark R, Whitfield GK, Haussler CA, Hsieh JC, Thompson PD, Selznick SH, Dominguez CE, Jurutka PW. "The nuclear vitamin D receptor: Biological and molecular regulatory properties revealed." *J. Bone Miner. Res* 13, 325–349 (1998).

Heaney, Robert P, Dowell MS, Hale CA, Bendich A. "Calcium absorption varies within the reference range for serum 25-hydroxyvitamin D." *J Am Coll Nutr* 22:142-6. (2003).

Heaney, Robert P. "Guidelines for optimizing design and analysis of clinical studies of nutrient effects." *Nutrition reviews*. 72(1):48±54 (2014)

Hendrickson, Whitney K, Flavin R, Kasperzyk JL, Fiorentino M, Fang F, Lis R, Fiore C, Penney KL, Ma J, Kantoff PW, Stampfer MJ, Loda M, Mucci LA, Giovannucci E. "Vitamin D receptor protein expression in tumor tissue and prostate cancer progression." *J. Clin. Oncol*. 29, 2378–2385 (2011).

Henry, Helen L and Norman AW. "Vitamin D: metabolism and biological actions." *Annu Rev Nutr*; 4:493–520 (1984).

Henry, Helen L. "Parathyroid hormone modulation of 25-hydroxyvitamin D3 metabolism by cultured chick kidney cells is mimicked and enhanced by forskolin." *Endocrinology* 116:503–10 (1985).

Höbaus, Julia, Fetahu I, Khorchide M, Manhardt T and Kallay E. "Epigenetic regulation of the 1,25-dihydroxyvitamin D3 24-hydroxylase (CYP24A1) in colon cancer cells." *J. Steroid Biochem. Mol. Biol.* 136, 296–299 (2013).

Höbaus, Julia, Thiem U, Hummel DM and Kallay E. "Role of calcium, vitamin d, and the extrarenal vitamin D hydroxylases in carcinogenesis." *Anticancer Agents Med. Chem.* 13, 20–35 (2013).

Holick, Micheal F. "Photosynthesis of vitamin D in the skin: effect of environment and life-style variables." *Fed Proc* 46:1876–1882 (1987).

Holick, Micheal F. "Noncalcemic actions of 1,25-dihydroxyvitamin D3 and clinical applications." *Bone* 17(2 suppl): 107S–111S (1995).

Holick, Micheal F, Siris ES, Binkley N, Beard MK, Khan A, Katzer JT, Petruschke RA, Chen E, de Papp AE. "Prevalence of vitamin D inadequacy among postmenopausal North American women receiving osteoporosis therapy." *J Clin Endocrinol Metab* 90:3215-24 (2005).

Holick, Micheal F. "Resurrection of vitamin D deficiency and rickets." *The Journal of clinical investigation* 116:2062–2072 (2006).

Holick, Micheal F. "The role of vitamin D for bone health and fracture prevention." *Curr Osteoporos Rep* 4:96–102 (2006).

Holick, Micheal F, Garabedian M. "Vitamin D: photobiology, metabolism, mechanism of action, and clinical applications. In: Favus MJ, ed. Primer on the metabolic bone diseases and disorders of mineral metabolism." 6th ed. Washington, DC: *American Society for Bone and Mineral Research* 129-37 (2006).

Holick, Micheal F, Binkley NC, Bischoff-Ferrari HA, Gordon CM, Hanley DA, Heaney RP, Murad MH, Weaver CM. "Evaluation, treatment, and prevention of vitamin D deficiency: an Endocrine Society clinical practice guideline." *J. Clin. Endocrinol. Metab.* 96, 1911–1930 (2011).

Horvath, Henrik C, Lakatos P, Kósa JP, Bácsi K, Borka K, Bises G, Nittke T, Hershberger PA, Speer G, Kállay E. "The candidate oncogene CYP24A1: a potential biomarker for colorectal tumorigenesis." *J. Histochem. Cytochem.* 58 277-285 (2010).

Hsu, Ju-Yu, Feldman D, McNeaL JE & Peehl DM. "Reduced 1α-hydroxylase activity in human prostate cancer cells correlates with decreased susceptibility to 25-hydroxyvitamin D3-induced growth inhibition." *Cancer Res.* 61, 2852–2856 (2001).

Hummel, Doris M, Fetahu IS, Groschel C, Manhardt T and Kallay E. "Role of proinflammatory cytokines on expression of vitamin D metabolism and target genes in colon cancer cells." *J. Steroid Biochem. Mol. Biol.* 144, 91–95 (2014).

Jones, Glenville, Prosser DE and Kaufmann M. "Cytochrome P450-mediated metabolism of vitamin D." *J. Lipid Res.* 55, 13–31 (2014).

Karin, Micheal and Lin A. "NF-kB at the crossroads of life and death." *Nat. Immunol.* 3:221–27 (2002).

Komagata, Sayaka., Nakajima M, Takagi S, Mohri T, Taniya T, Yokoi T. "Human CYP24 catalyzing the inactivation of calcitriol is post-transcriptionally regulated by miR-125b." *Mol. Pharmacol.* 76, 702–709 (2009).

Krishnan, Aruna V, Moreno J, Nonn L, Swami S, PeehlDM, FeldmanD. Calcitriol as a chemopreventive and therapeutic agent in prostate cancer: role of anti-inflammatory activity. *J. Bone Miner. Res.* 22(Suppl.2): V74–80 (2007).

Krishnan, Aruna V and Feldman D. "Molecular pathways mediating the anti-inflammatory effects of calcitriol: implications for prostate cancer chemoprevention and treatment." *Endocr. Relat. Cancer* 17, R19–R38 (2010).

Krishnan, Aruna V, Peng L, Wang J, Moreno J, Feldman D. "Tissue-selective regulation of aromatase expression by calcitriol: implications for breast cancer therapy." *Endocrinology* 151, 32–42 (2010).

Lappe, Joan M, Travers-Gustafson D, Davies KM, Recker RR and Heaney RP. "Vitamin D and calcium supplementation reduces cancer risk: results of a randomized trial." *Am. J. Clin. Nutr.* 85, 1586–1591 (2007).

Lappe, Joan, Watson P, Travers-Gustafson D, Recker R, Garland C, Gorham E, Baggerly K, McDonnell S. "Effect of Vitamin D and Calcium Supplementation on Cancer Incidence in Older Women: A Randomized Clinical Trial." *JAMA: The Journal of the American Medical Association* 317(12):1234±43 (2017).

Liel, Yair, Shany S, Smirnoff P and Schwartz B. "Estrogen increases 1,25 dihydroxyvitamin D receptors expression and bioresponse in the rat duodenal mucosa." *Endocrinology* 140, 280–285 (1999).

Liu, Min, Lee MH, Cohen M, Bommakanti M, Freedman LP. "Transcriptional activation of the Cdk inhibitor p21 by vitamin D3 leads to the induced differentiation of the myelomonocytic cell line U937." *Genes Dev*. 10:142–53 (1996).

Lopes, Nair, Sousa B, Martins D, Gomes M, Vieira D, Veronese LA, Milanezi F, Paredes J, Costa JL, Schmitt F. "Alterations in vitamin D signalling and metabolic pathways in breast cancer progression: a study of VDR, CYP27B1 and CYP24A1 expression in benign and malignant breast lesions." *BMC Cancer* 10, 483 (2010).

Luo, Wei, Karpf AR, Deeb KK, Muindi JR, Morrison CD, Johnson CS, Trump DL. "Epigenetic regulation of vitamin D 24-hydroxylase/ CYP24A1 in human prostate cancer." *Cancer Res.* 70, 5953–5962 (2010).

Ly, Lan H, Zhao XY, Holloway L and Feldman D. "Liarozole acts synergistically with 1α, 25-dihydroxyvitamin D3 to inhibit growth of DU145 human prostate cancer cells by blocking 24-hydroxylase activity." *Endocrinology* 140, 2071–2076 (1999).

Maier, Sandra, Daroqui MC, Scherer S, Roepcke S, Velcich A, Shenoy SM, Singer RH, Augenlicht LH. "Butyrate and vitamin D3 induce D3 induce transcriptional attenuation at the cyclin transcriptional attenuation at the cyclin D1 locus in colonic carcinoma cells." *J. Cell. Physiol*. 218, 638–642 (2009).

Manson, JoAnn E, Bassuk SS, Lee IM, Cook NR, Albert MA, Gordon D, Zaharris E, Macfadyen JG, Danielson E, Lin J, Zhang SM, Buring JE. "The VITamin D and OmegA-3 TriaL (VITAL): rationale and design of a large randomized controlled trial of vitamin D and marine omega-3 fatty acid supplements for the primary prevention of cancer and cardiovascular disease." *Contemp. Clin. Trials* 33, 159–171 (2012).

Mantell, Donald J, Owens PE, Bundred NJ, Mawer EB, Canfield AE. "1a,25-Dihydroxyvitamin D3 inhibits angiogenesis in vitro and in vivo." *Circ. Res.* 87:214–20 (2000).

Mantovani, Alberto, Allavena P, Sica A. & Balkwill F. "Cancer-related inflammation." *Nature* 454, 436–444 (2008).

Marik, Radharani, Fackler M, Gabrielson E, Zeiger MA, Sukumar S, Stearns V, Umbricht CB. "DNA methylation-related vitamin D receptor insensitivity in breast cancer." *Cancer Biol. Ther.* 10, 44–53 (2010).

Martin, Aline, David V and Quarles LD. "Regulation and function of the FGF23/klotho endocrine pathways." *Physiol. Rev.* 92, 131–155 (2012).

Maruyama, Reo, Aoki F, Toyota M, Sasaki Y, Akashi H, Mita H, Suzuki H, Akino K, Ohe-Toyota M, Maruyama Y, Tatsumi H, Imai K, Shinomura Y, Tokino T. "Comparative genome analysis identifies the vitamin D receptor gene as a direct target of p53-mediated transcriptional activation." *Cancer Res.* 66, 4574–4583 (2006).

McKenna, Malachi J. "Differences in vitamin D status between countries in young adults and the elderly." *Am J Med* 93(1):69–77 (1992).

Miller, Gary J, Stapleton GE, Hedlund TE and Moffat KA. "Vitamin D receptor expression, 24-hydroxylase activity, and inhibition of growth by 1α, 25-dihydroxyvitamin D3 in seven human prostatic carcinoma cell lines." *Clin. Cancer Res.* 1, 997–1003 (1995).

Miller, Walter L and Auchus RJ. "The molecular biology, biochemistry, and physiology of human steroidogenesis and its disorders." *Endocr. Rev.* 32, 81–151 (2011).

Morelli, Susana, Buitrago C, Boland R and de Boland AR. "The stimulation of MAPk inaseby 1,25(OH)(2)-vitaminD3 in skeletal muscle cells is mediated by protein kinase C and calcium." *Mol. Cell. Endocrinol.* 173, 41–52 (2001).

Moreno, Jacqueline, Krishnan AV, Swami S, Nonn L, Peehl DM, Feldman D. "Regulation of prostaglandin metabolism by calcitriol attenuates growth stimulation in prostate cancer cells." *Cancer Res.* 65, 7917–7925 (2005).

Moukayed, Meis and Grant WB. "Molecular link between vitamin d and cancer prevention." *Nutrients* 5, 393-4021 (2013).

Muindi, Josephia R, Yu WD, Ma Y, Engler KL, Kong RX, Trump DL, Johnson CS. "CYP24A1 inhibition enhances the antitumor activity of calcitriol." *Endocrinology* 151, 4301–4312 (2010).

Murayama, Akiko, Kim MS, Yanagisawa J, Takeyama K, Kato S. "Transrepression by a liganded nuclear receptor via a bHLH activator through co-regulator switching." *EMBO J* 23:1598–608 (2000).

Murayama, Akiko, Takeyama K, Kitanaka S, Kodera Y, Kawaguchi Y, Hosoya T, Kato S. "Positive and negative regulations of the renal 25-hydroxyvitamin D3 1alpha-hydroxylase gene by parathyroid hormone, calcitonin, and 1alpha,25(OH)2D3 in intact animals." *Endocrinology* 140:2224–31 (1999).

National Academy of Sciences, Institute of Medicine, Food and Nutrition Board. Dietary Reference Intakes for Calcium, phosphorus, Magnesium, Vitamin D, and Fluoride. Washington, DC: *National Academy Press* (1997).

Nemere, Ilka, Farach-Carson MC, Rohe B, Sterling TM, Norman AW, Boyan BD, and Safford SE. "Ribozyme knockdown functionally links a 1,25(OH)2D3membrane binding protein(1,25D3-MARRS) and phosphate uptake in intestinal cells." *Proc. Natl. Acad. Sci. U.S.A.* 101, 7392–7397 (2004).

Nonn, Larisa, Peng L, Feldman D and Peehl DM. "Inhibition of p38 by vitamin D reduces interleukin-6 production in normal prostate cells via mitogen- activated protein kinase phosphatase 5: implications for prostate cancer prevention by vitamin D." *Cancer Res.* 66, 4516–4524 (2006).

Norman, Anthony W. "Vitamin D receptor: new assignments for an already busy receptor." *Endocrinology* 147, 5542–5548 (2006).

O'Kelly, James, Morosetti R, Koeffler HP. "Vitamin D and hematological malignancy." *Vitamin D*, Volume II, pp. 1727–40 (2005).

Omdahl, John L, Morris HA, May BK. "Hydroxylase enzymes of the vitamin D pathway: expression, function, and regulation." *Annu Rev Nutr* 22:139–66 (2002).

Padi, Sathish KR, Zhang Q, Rustum YM, Morrison C and Guo B. "MicroRNA-627 mediates the epigenetic mechanisms of vitamin D to suppress proliferation of human colorectal cancer cells and growth of xenograft tumors in mice." *Gastroenterology* 145, 437–446 (2013).

Palayoor, Sanjeewani T, Youmell MY, Calderwood SK, Coleman CN, Price BD. "Constitutive activation of IκB kinase αand NF-κB in prostate cancer cells is inhibited by ibuprofen." *Oncogene* 18:7389–94 (1999).

Palmer, Hector G, Larriba MJ, García JM, Ordóñez-Morán P, Peña C, Peiró S, Puig I, Rodríguez R, de la Fuente R, Bernad A, Pollán M, Bonilla F, Gamallo C, de Herreros AG, Muñoz A. "The transcription factor SNAIL represses vitamin D receptor expression and responsiveness in human colon cancer." *Nat. Med.* 10, 917–919 (2004).

Park, Jae-Il, Lee MG, Cho K, Park BJ, Chae KS, Byun DS, Ryu BK, Park YK, Chi SG. "Transforming growth factor-ß1 activates interleukin-6 expression in prostate cancer cells through the synergistic collaboration of the Smad2, p38-NF-KB, JNK, and Ras signaling pathways." *Oncogene* 22:4314–32(2003).

Peehl, Donna M, Shinghal R, Nonn L, Seto E, Krishnan AV, Brooks JD, Feldman D. "Molecular activity of 1,25-dihydroxyvitaminD3 in primary cultures of human prostatic epithelial cells revealed by cDNA microarray analysis." *J. Steroid Biochem. Mol. Biol.* 92:131–41 (2004).

Peng, Lihong, Malloy PJ, Feldman D. "Identification of a functional vitamin D response element in the human insulin-like growth factor binding protein-3 promoter." *Mol. Endocrinol.* 18:1109–19 (2004).

Pike, J Wesley and Meyer MB. "Regulation of mouse Cyp24a1 expression via promoter-proximal and downstream-distal enhancers highlights new concepts of 1,25-dihydroxyvitamin D3 action." *Archives of biochemistry and biophysics* 523:2–8 (2012).

Pike, J Wesley, Meyer MB and Bishop KA. "Regulation of target gene expression by the vitamin D receptor - an update on mechanisms." *Rev. Endocr. Metab. Disord.* 13, 45–55 (2012).

Ponchon, Gerard, DeLuca HF. "The role of the liver in the metabolism of vitamin D." *The Journal of clinical investigation.* 48:1273–1279 (1969).

Rose-John, Stefan and Schooltink H. "Cytokines are a therapeutic target for the prevention of inflammation induced cancers." *Recent Results Cancer Res.* 174:57–66 (2007).

Sahin, Mehmet, Sahin E, Gumuslu S. "Cyclooxygenase-2 in cancer and angiogenesis." *Angiology* 60:242–53 (2009).

Sakamoto, Shinichi, Ryan AJ, Kyprianou N. "Targeting vasculature in urologic tumors: mechanistic and therapeutic significance." *J. Cell. Biochem.* 103:691–708 (2008).

Samuel, Sam and Sitrin MD. "Vitamin D's role in cell proliferation and differentiation." *Nutr. Rev.* 66, S116–S124 (2008).

Shimada, Takashi, Kakitani M, Yamazaki Y, Hasegawa H, Takeuchi Y, Fujita T, Fukumoto S, Tomizuka K, Yamashita T. "Targeted ablation of Fgf23 demonstrates an essential physiological role of FGF23 in phosphate and vitamin D metabolism." *J Clin Invest* 113:561–8 (2004).

Skowronski, Roman J, Peehl DM and Feldman D. "Vitamin D and prostate cancer: 1,25 dihydroxyvitamin D3 receptors and actions in human prostate cancer cell lines." *Endocrinology* 132, 1952–1960 (1993).

Slominski, Andrzej T, Semak I, Zjawiony J, Wortsman J, Li W, Szczesniewski A, Tuckey RC. "The cytochrome P450scc system opens an alternate pathway of vitamin D3 metabolism." *FEBS J.* 272, 4080–4090 (2005).

Slominski, Andrzej T, Kim TK, Li W, Yi AK, Postlethwaite A, Tuckey RC. "The role of CYP11A1 in the production of vitamin D metabolites and their role in the regulation of epidermal functions." *J. Steroid Biochem. Mol. Biol.* 144(Pt A), 28–39 (2014).

Slominski, Andrzej T, Kim TK, Hobrath JV, Oak ASW, Tang EKY, Tieu EW, Li W, Tuckey RC, Jetten AM. "Endogenously produced nonclassical vitamin D hydroxy-metabolites act as "biased" agonists on

VDR and inverse agonists on RORα and RORγ." *J. Steroid Biochem. Mol. Biol.* 173, 42–56 (2017).

Slominski, Andrzej T, Brożyna AA, Skobowiat C, Zmijewski MA, Kim TK, Janjetovic Z, Oak AS, Jozwicki W, Jetten AM, Mason RS, Elmets C, Li W, Hoffman RM, Tuckey RC. "On the role of classical and novel forms of vitamin D in melanoma progression and management." *J. Steroid Biochem. Mol. Biol.* (2017).

Srinivasan, Malini, Parwani AV, Hershberger PA, Lenzner DE and Weissfeld JL. "Nuclear vitamin D receptor expression is associated with improved survival in non-small cell lung cancer." *J. Steroid Biochem. Mol. Biol.* 123, 30–36 (2011).

Stambolsky, Perry, Tabach Y, Fontemaggi G, Weisz L, Maor-Aloni R, Shiff I, Kogan I, Shay M, Kalo E, Blandino G, Simon I, Oren M, Rotten V. et al. "Modulation of the vitamin D3 response by cancer associated mutant p53." *Cancer Cell.* 17, 273–p285 (2010).

Swami, Srilatha, Krishnan AV, Peehl DM and Feldman D. "Genistein potentiates the growth inhibitory effects of 1,25-dihydroxyvitamin D(3) in DU145 human prostate cancer cells: role of the direct inhibition of CYP24 enzyme activity." *Mol. Cell Endocrinol.* 12, 12 (2005).

Tongkao-On, Wannit, Carter S, Reeve VE, Dixon KM, Gordon-Thomson C, Halliday GM, Tuckey RC, Mason RS. "CYP11A1 in skin: an alternative route to photoprotection by vitamin D compounds." *J. Steroid Biochem. Mol. Biol.* 148, 72–78 (2015).

Vuolo, Laura, Di Somma C, Faggiano A, Colao A. "Vitamin D and cancer." *Front Endocrinol (Lausanne)* 3:58 (2012).

Wang, Jennifer Y, Swami S, Krishnan AV and Feldman D. "Combination of calcitriol and dietary soy exhibits enhanced anticancer activity and increased hypercalcemic toxicity in a mouse xenograft model of prostate cancer." *Prostate* 72, 1628–1637 (2012).

Wang, Lilin, Flanagan JN, Whitlatch LW, Jamieson DP, Holick MF, Chen TC. "Regulation of 25-hydroxyvitamin D-1a-hydroxylase by epidermal growth factor in prostate cells." *J. Steroid Biochem. Mol. Biol.* 89–90, 127–130 (2004).

Wang, Qingsong, He Y, Shen Y, Zhang Q, Chen D, Zuo C, Qin J, Wang H, Wang J, Yu Y. "Vitamin D inhibits COX-2 expression and inflammator response by targeting thioesterase superfamily member 4." *J. Biol. Chem.* 289, 11681–11694 (2014).

Wang, Yongji, Zhu J and DeLuca HF. "Where is the vitamin D receptor?" *Arch. Biochem. Biophys.* 523, 123–133 (2012).

Webb, Ann R, Pilbeam C, Hanofin N, Holick M. "An evaluation of the relative contributions of exposure to sunlight and of diet to the circulating concentrations of 25-hydroxyvitamin D in an elderly nursing home population in Boston." *Am J Clin Nutr.* 51:1075–1081 (1990).

Welsh, JoEllen. "Cellular and molecular effects of vitamin D on carcinogenesis." *Arch. Biochem. Biophys.* 523, 107–114 (2012).

Whitlatch, Lyman W, Young MV, Schwartz GG, Flanagan JN, Burnstein KL, Lokeshwar BL, Rich ES, Holick MF, Chen TC. "25-Hydroxyvitamin D-1α-hydroxylase activity is diminished in human prostate cancer cells and is enhanced by gene transfer." *J. Steroid Biochem. Mol. Biol.* 81, 135–140 (2002).

Woo, Tony Choon Seng, Choo R, Jamieson M, Chander S, Vieth R. "Pilot study: potential role of vitamin D (cholecalciferol) in patients with PSA relapse after definitive therapy." *Nutr. Cancer* 51:32–36 (2005).

Yates, Luke A, Norbury CJ and Gilbert RJ. "The long and short of microRNA." *Cell* 153, 516–519 (2013).

Young, Micheal V, Schwartz GG, Wang L, Jamieso DP, Whitlatch LW, Flanagan JN, Lokeshwar BL, Holick MK and Chen TC. "The prostate 25-hydroxyvitamin D-1a-hydroxylase is not influenced by parathyroid hormone and calcium: implications for prostate cancer chemoprevention by vitamin D." *Carcinogenesis* 25, 967–971 (2004).

Zanatta, Leila, Zamoner A, Zanatta AP, Bouraima-Lelong H, Delalande C, Bois C, Carreau S, Silva FR. "Nongenomic and genomic effects of 1alpha,25(OH)2 vitamin D3 in rat testis." *Life Sci.* 89, 515–523(2011).

About the Editor

Curriculum Vitae Et Studiorum
Prof. Vittorio Unfer

Vittorio Unfer, M.D., was born in Rome on October 12, 1963 where he currently lives and works as an Obstetrics and Gynaecology Specialist, researcher and university lecturer.

He graduated with honors in 1991 in Medicine and Surgery at the Faculty of Medicine of Sapienza University of Rome, successively he decided to enroll to the specialization in obstetrics and gynecology at the II Clinical Institute of Obstetrics and Gynaecology of Policlinico Umberto I in Rome, obtaining a score of 70 out of 70 with honors in 1995.

Academic activities

At present, Vittorio Unfer is Lecturer at the Department of Experimental Medicine of the Sapienza University of Rome.

He was Full Professor at the University IPUS (Advanced Training Institute) of Chiasso, Switzerland and a Master Professor in "Unconventional Medicine in assisting Obstetrics and Nursing Autonomy and midwife in the management of pregnancy, childbirth and postpartum low risk" at the Faculty of Medicine of the University of Modena and Reggio Emilia..

In 2010 he was appointed as a lecturer in Endocrinology childhood, adolescence and women at the University Master's Degree, School of Specialization in Endocrinology, Degree Course in Medicine and Surgery of University of Messina.

From 2008 to 2011 he teaches nutrition and food disease in pregnancy at the School of Specialization in Gynaecology and Obstetrics, Degree Course in Obstetrics, University of Perugia.

In the Academic Year 2001-2002 is a Professor of Health Statistics at the Department of Gynaecological Sciences, Perinatal and Child-care, Degree Course in Obstetrics, Sapienza University of Rome.

From 1999 to 2000 he works as a Professor of Forensic Medicine at the Department of Gynaecological Sciences, Perinatal and Child-care, Degree Course in Obstetrics, Sapienza University of Rome.

Research activities

He is also the author of numerous publications: over 50 international journals, more than 90 publications in national magazines, over 10 monographs and edited works, more than 100 abstracts in national and international periodicals, among the most important and recent is relevant to mention:

1. Short-term effects of metformin and myo-inositol in women with polycystic ovarian syndrome (PCOS): a meta-analysis of randomized clinical trials.
   Facchinetti F, Orrù B, Grandi G, Unfer V. Gynecol Endocrinol. 2019 Mar;35(3):198-206. doi: 10.1080/09513590.2018.1540578.
2. Vitamin D: a steroid hormone with progesterone-like activity.
   Monastra G, De Grazia S, De Luca L, Vittorio S, Unfer V. Eur Rev Med Pharmacol Sci. 2018 doi: 10.26355/eurrev_201804_14845.
3. Myo-inositol effects in women with PCOS: a meta-analysis of randomized controlled trials.
   Unfer V, Facchinetti F, Orrù B, Giordani B, Nestler J. Endocr Connect. 2017 Nov;6(8):647-658.
4. Myo-inositol: from induction of ovulation to menopausal disorder management.
   Gullo G, Carlomagno G, Unfer V, D'Anna R. Minerva Ginecol. 2015 Oct;
5. Second trimester amniotic fluid myo-inositol concentrations in women later developing gestational diabetes mellitus or pregnancy-induced hypertension.
   Santamaria A, Corrado F, Baviera G, Carlomagno G, Unfer V, D'anna R.. J Matern Fetal Neonatal Med. Epub 2015 Sep 12;
6. Reflections on inositol(s) for PCOS therapy: steps toward success.
   Nestler JE, Unfer V. Gynecol Endocrinol. 2015 Jul;
7. Peripheral neuropathy in obstetrics: efficacy and safety of α-lipoic acid supplementation.

Costantino M, Guaraldi C, Costantino D, De Grazia S, Unfer V. Eur Rev Med Pharmacol Sci. 2014;18(18):2766-71. Review;

8. The rationale of the myo-inositol and D-chiro-inositol combined treatment for polycystic ovary syndrome.
   Dinicola S, Chiu TT, Unfer V, Carlomagno G, Bizzarri M. J Clin Pharmacol. 2014 Oct;54(10):1079-92. doi: 10.1002/jcph.362. Epub 2014 Jul 18. Review;
9. Myo-inositol in Down syndrome amniotic fluid. A case-control study.
   Santamaria A, Corrado F, Interdonato ML, Baviera G, Carlomagno G, Cavalli P, Unfer V, D'Anna R.
   Prenat Diagn. 2014 Sep;34(9):917-8. doi: 10.1002/pd.4398. Epub 2014 Jul 5;
10. Updates on the myo-inositol plus D-chiro-inositol combined therapy in polycystic ovary syndrome.
    Unfer V, Porcaro G. Expert Rev Clin Pharmacol. 2014 Sep;7(5):623-31. doi: 10.1586/17512433.2014.925795. Epub 2014 Jun 5. Review;
11. Hyperinsulinemia Alters Myo-inositol to d-chiroinositol Ratio in the Follicular Fluid of Patients With PCOS.
    Unfer V, Carlomagno G, Papaleo E, Vailati S, Candiani M, Baillargeon JP. Reprod Sci. 2014 Feb 4;21(7):854-858. [Epub ahead of print];

In 2005 he contributed to the drafting of the Treaty of Gynaecology and Obstetrics published by Verduci and currently is still an active member of Verduci ECM Scientific Committee.

The intensive research in recent times of Professor Vittorio Unfer is connected to aspects of nutrition and use of food supplements in obstetrics and gynaecology clinical practice.

In the academic sector Professor Vittorio Unfer participated in many other important scientific research focused on the morphofunctional study of uterine contractile activity with previous uterine scar, the triggering mechanisms of labor, the treatment of the threatened miscarriage abortion and prevention of preterm labor in high-risk patients, the treatment of hypogonadotropic male hypogonadism with FSH, and several in-depth and comparative experimentations on the fetal maturity.

It is few years that the scientific contribution of Professor Vittorio Unfer is related to the investigations focused on PCOS (polycystic ovary syndrome) and in particular on the role of second messengers in ovarian function.

Prof. Vittorio Unfer resulted to be the first promoter of the use of myo-inositol in the treatment for polycystic ovary syndrome; his in-depth research pointed out the beneficial effects of the myo-inositol molecule on ovarian function and metabolic parameters.

These results were highly appreciated by the international community due to the findings of Professor Vittorio Unfer. Today the treatment with myo-inositol for women with PCOS is adopted in more than 36 countries.

Private activities

Prof. Vittorio Unfer currently is the Medical Director and Gynaecologist at "A.G.UN.CO. – Studi Polimedici Specialistici".

He is in charge as President of the Pharmaceutical Society Lo.Li Pharma Ltd, Chairman of the Board of Lo.Li Pharma International Ltd. As well as Chairman of the Board of Farmares Ltd, research and development company.

Prof. Vittorio Unfer, moreover holds the position of Vice President of SIFIPsi (Italian Society of Phytotherapy and supplements in Psychopathology), he is member of the Pharmaceutical Society Bioitalia Ltd, Co- founder of ISCHOM (International Society of Chocolate, Cocoa in Medicine) and is Past-President of SIFIOG, Italian Society of Phytotherapy and supplements in Obstetrics and Gynaecology.

Since 1995 he attended more than 25 training courses to enhance his professional skills.

He participated as a speaker in many national and international conferences in the Endocrinology Gynaecology and Obstetrics fields.

Lately Prof. Vittorio Unfer has been guest on various TV shows and radio broadcasting programs as expert on issues related to infertility and

PCOS, such as "Sulle Ali della Salute", " Medicina, Salute e Benessere", "Mattino Cinque" on Channel 5, Tg Norba 24 and Sky Channel 913.

"In compliance with the Italian legislative Decree no. 196 dated 30/06/2003, I hereby authorize you to use and process my personal details contained in this document".

INDEX

#

β

A

B

C

D

E

F

G

H

I

N

O

P

Q

R

S

T

U

V

X